AF581771

Bioelectric Sensors

Bioelectric Sensors

Editor

Spyridon Kintzios

MDPI • Basel • Beijing • Wuhan • Barcelona • Belgrade • Manchester • Tokyo • Cluj • Tianjin

Editor
Spyridon Kintzios
Agricultural University of
Athens
Greece

Editorial Office
MDPI
St. Alban-Anlage 66
4052 Basel, Switzerland

This is a reprint of articles from the Special Issue published online in the open access journal *Biosensors* (ISSN 2079-6374) (available at: https://www.mdpi.com/journal/biosensors/special_issues/bioelectric_sens).

For citation purposes, cite each article independently as indicated on the article page online and as indicated below:

LastName, A.A.; LastName, B.B.; LastName, C.C. Article Title. *Journal Name* **Year**, *Article Number*, Page Range.

ISBN 978-3-03943-084-0 (Hbk)
ISBN 978-3-03943-085-7 (PDF)

Contents

About the Editor

Spyridon Kintzios holds a Ph.D. in Genetics (TU Munich), with background studies in Plant & Agricultural Science and Chemistry, as well as a DAAD scholarship at the Max-Planck Institute for Biochemistry in Munich. He is Full Professor and the current Rector of the Agricultural University of Athens. He has over 30 years of work experience in Biotechnology, particularly in the fields of Biosensors and Cell Biology. He is the author or co-author of 120 peer-reviewed articles in cited international scientific journals, as well as the author or co-author of more than 90 international book chapters and conference presentations. He is also the editor of five international books and holder of two European and two national patents.

Preface to "Bioelectric Sensors"

Bioelectric sensors represent a continuously growing segment of biosensor technologies and applications, offering key advantages in terms of their practicability of application and scalability of manufacturing. Progress in related fields has been considerably boosted by advances in microelectronics and nanotechnology, in general, while the quiver of biocompatible materials serving as intermediates between biorecognition elements and electronic components has been impressively expanded. Although bioelectric sensors share many traits with electrochemical sensors, especially regarding common features of instrumentation, they are focused on the measurement of the electric properties of biorecognition elements as a reflection of cellular, biological, and biomolecular functions in a rapid, very sensitive, and often non-invasive manner. Bioelectric sensors offer a plethora of options in terms both of assay targets (molecules, cells, organs, and organisms) and methodological approaches (e.g., potentiometry, impedance spectrometry, patch-clamp electrophysiology). Irrespective of the method of choice, "bioelectric profiling" is being rapidly established as a superior concept for several applications, including in vitro toxicity, signal transduction, real-time medical diagnostics, environmental risk assessment, and drug development.

The Special Issue *Bioelectric Sensors* is the first one exclusively dedicated to advanced and emerging concepts and technologies of bioelectric sensors. Contributed articles focus on key topics such as reflectance-based pulse sensors, wireless monitoring systems, and bioelectric biochips and applications including non-invasive wound healing, cancer cell fingerprinting, differential drug screening, and advanced pacemaker performance modeling. All approaches are handled in the context of point-of-care/portable and wireless instrumentation and intelligent bioelectric sensing platforms.

Spyridon Kintzios
Editor

Editorial

Bioelectric Sensors: On the Road for the 4.0 Diagnostics and Biomedtech Revolution

Spyridon Kintzios

Laboratory of Cell Technology, Faculty of Biotechnology, Agricultural University of Athens/EU-CONEXUS European University, 11855 Athens, Greece; skin@aua.gr

Received: 30 July 2020; Accepted: 4 August 2020; Published: 11 August 2020

Bioelectric sensors lie, by definition, on the interface between biological elements and electronic circuits, irrespective of scale, manufacturing method, and working principle. They distinguish themselves from electrochemical sensors in the sense that they rely exclusively on cells, tissues, and even organs as the biorecognition elements, instead of using only biomolecular moieties, such as antibodies, enzymes, or oligonucleotides.

Bioelectric sensors are quite popular as tools for rapidly accessing the cellular physiologic status: this is a field where both potentiometric and bioimpedance-based biosensors are being increasingly used for toxicity and/or metabolic effects screening [1–4]. Recent examples in the later application area are represented both by the XF Extracellular Flux Analyzer platform for metabolic assays [5,6] and the Cell Culture Metabolite Biosensor prototype [7] for measuring glycolytic metabolism and inhibitor effects on CD4+ T cells. More advanced systems and approaches are able to provide considerable volumes of experimental information, for example, by means of impedance frequency spectrometry, which can, in turn, be used to train dedicated software for identification and classification of data subgroups. On the biological side, significant progress has been made by immobilizing cells either two-dimensionally onto the surface of conducting electrodes or in a three-dimensional configuration in the appropriate gel: the last option usually contributes to significant simplification and increased efficiency of operation, as well as extended cell viability and storage stability [8].

Among the advantageous traits of bioelectric sensors, speed, non-invasiveness and low cost per assay are the most prominent ones. As a paradigm, bioelectric profiling toxicity assays against pesticide residues can be conducted within a few minutes whereas conventional enzyme-based optical assays may require several hours or even days [9,10]. On the downside, information on the electric properties of living cells and tissues is rarely associated with specific molecular functions, unless the cellular biorecognition element is tailor-made to couple certain biochemical responses to a bioelectric mechanism. Such is the case of membrane-engineered cells and cells with synthetic gene circuits [11–13]. Otherwise, the preferable field of application for bioelectric sensors remains that of a more holistic screening of cellular physiology, in particular cell toxicity and membrane channel activity.

Similar to electrochemical sensors but also distinct from them, bioelectric sensors are able to monitor in real-time, often continuously, physiological patterns and transmit results via Bluetooth/internet to remote data storage, process, and interpretation sites. In several cases, monitoring is conducted non-invasively and, most importantly, not requiring sample extraction. In this way, it is possible to couple biosensors with dedicated, true point-of-care (POC) or point-of-test (POT) platforms (e.g., wearables) that are integrated in various Internet of Things (IoT) networks, including smartphone-based telemetry and e-health applications [14–19].

In this context, the present Special Issue is not only the first volume exclusively dedicated to bioelectric sensors. In a genuinely emblematic approach, its seven articles, selected through very rigorous peer review and authored by experts of the highest caliber globally, deal with the foremost and advanced technologies and applications in the field of bioelectric sensors. Moreover, they focus on system integration to deliver practical point-of-care/portable and wireless instrumentation and intelligent bioelectric sensing platforms. These will be presented in more detail in following.

Organic biosensors with minimum power consumption represent the next stage of pulse meters, i.e., devices serving as non-invasive rapid medical diagnostic tools by measuring the rate of rhythmic contraction and expansion of an artery, in sync with the heart. They are based on the photoplethysmogram (PPG) principle, according to which, changes in reflected light, detectable as a PPG signal, correspond to changes in the volume of the underlying artery. In their contribution, Elsamnah et al. [20] report the development of a novel organic optoelectronic device purposed as a pulse oximeter and based on two alternative designs using large organic photodiodes (OPDs) and organic light-emitting diodes (OLEDs). These two models were simulated by representing the simplified four-layer structure of a finger model, with red OLED being preferred over green and infrared ones. Both devices were reliable and obtained a clear and stable PPG signal from a healthy individual, with minimum power consumption in wireless monitoring of PPG waveforms. The biosensor pulse meter showed promising results with ultra-low power consumption, 8 μW at 18 dB signal-to-noise ratio (SNR), and demonstrated its ability to measure a clear PPG signal up to 46 dB SNR at a constant current of 93.6 μA. Coupled with a low manufacturing cost, the novel system is very promising for long-term wireless PPG signal monitoring, possibly also as part of a wearable medical device.

Next, Kiani et al. [21] report on the combination of a miniaturized—and therefore fully portable—p-n junction-based Si biochip with impedance spectroscopy, and using the industrial metal-binding, metal-remediating bacteria *Lysinibacillus sphaericus* JG-A12 as the biosensing element. The ohmic or Schottky contacts in the biochip was modelled as the combination of resistors and capacitors, while impedance spectrometry was modelled by using constant phase elements (CPEs). The bulk capacitance of the depletion region of the semiconductor and the capacitance of the Schottky contacts between electrodes and semiconductor contributed to the impedance spectra of the biochips. A linear pattern of response was determined with increasing bacteria concentration measured at test frequencies of 40 Hz, 400 Hz, and 4000 Hz.

Nanotechnology is a major accelerator in the race for continuous device miniaturization and, naturally, bioelectric sensors could not be kept out of this progress. Janssen et al. [22] elaborate on the use of carbon nanotubes (CNTs) for improved sensitivity and response time as potential candidates for PoC protein detection, with the detection of bovine serum albumin (BSA) as a proof-of-concept application. Having a nanometer-scale diameter, CNTs are characterized by large surface and high electrical conductivity, which renders them ideal substrates for manufacturing bioelectric sensing element at the nanoscale. When considering a large, three-dimensional population of such conductive nanoelements interacting with biological moieties—such as antibodies—the systemic conductivity depends on the topological alignment of the nanoelements in this network. In other words, depending on the interaction between nanoelements such as CNTs and their immediate environment, including antibodies and target antigens (analytes in a sample), any disruption of the network continuity will result in a measurable increase in the network's electrical resistance. This effect is called electrical percolation. The authors applied this working principle to develop a CNT-based, bioelectrical percolation sensor for rapid (10 min) BSA determination with a limit of detection of 2.89 ng/mL and a linear response between 5 and 45 ng/mL. The biosensor was built upon a disposable cellulose paper strip impregnated with CNTs and antibodies for protein detection, the electrical resistance of which was measured with a programmed Arduino Uno.

Application wise, one of the most intriguing and, at the same time, fascinating areas is the intercalation of bioelectric sensors with bioelectromagnetic medical interventions. Wound healing with the aid of external electric fields is such a case. Electrical simulation (ES) is one of the current electromagnetic therapeutic approaches to non-surgical wound healing. Lagoumintzis et al. [23] report on wireless micro current stimulation (WMCS), an alternative non-invasive and non-contact method to electrode-based ES. This approach utilizes the current-carrying capacity of charged air gas, based on the ability of nitrogen (N_2) and/or oxygen (O_2) molecules to accept or donate electrons, in order to distribute currents and voltages within the subject tissue. The authors applied an O_2^--induced microcurrent of 1.5–4.0 μA intensity in the patient's body by using a device capable of producing a specific number of charged particles which covered the wound area from a distance of 12–15 cm. Clinical observations after a three-month treatment period demonstrated the considerable reduction of massive pressure ulcers and the formation of healthy new epithelial tissue. Immunohistochemical analysis revealed both a suppression of inflammation upon WMCS treatment, as well as an increase in myofibroblastic activity, collagen formation, mast cell existence, and a reduced granulocyte aggregation. In essence, the application of tandem WMCS sessions led to reverse the wound-associated electrical leak that short-circuits the skin and to restore the physiological electric fields and ionic currents of the affected tissues. The potential benefits of wide adoption of WMCS in clinical practice as a non-invasive, reagent-free method for wound healing is more than obvious.

Bioelectric profiling is being rapidly established as a superior concept for several applications, including in vitro toxicity, signal transduction, real-time medical diagnostics, environmental risk assessment, and drug development. In the case of cancer, research in the field of hypoxia revealed how critical the pericellular oxygenation in a cell culture is. In this context, a critical marker for the monitoring the differentiation of cancer cells within a cell population is superoxide anion, which is mainly generated as a by-product of the oxidative phosphorylation by the electron transport chain of the mitochondria, is released to the mitochondrial matrix, where it is converted immediately to hydrogen peroxide. Mitochondrial hydrogen peroxide can then diffuse to the cytosol and the nucleus and react with other free radical species, alter signaling pathways or cause cellular damage. Along with other free radical species, superoxide has been found to mediate the development and/or survival of cancer cells and tumors, both in vivo and in vitro [24–26]. While hypoxia-regulated processes can result in the bad prognosis of conventional chemotherapy it is essential to monitor and control the cellular microenvironment. Mavrikou et al. [27] demonstrate an innovative and technologically disruptive approach for cell culture monitoring that can be used as an indicator for the response to different chemotherapy options. In particular, they investigated the accumulation of superoxide ions in cultured HeLa cervical cancer cells in response to different 5-fluorouracil (5-FU) concentrations. The anticancer activity of 5-FU emerges from the inhibition of thymidylate synthase (TS) activity during the S phase of the cell cycle and its incorporation into RNA and DNA of tumor cells, as well as from the generation mitochondrial ROS in the p53-dependent pathway [28–32]. Superoxide ion accumulation was monitored with the aid of an advanced bioelectric sensor based on Vero cells which were membrane-engineered with superoxide dismutase. As proven in several reports, the membrane potential of membrane-engineered Vero cell fibroblasts is affected by the interactions of electroinserted SOD molecules and superoxide anions, producing measurable changes in the membrane potential and can be used to determine superoxide extracellular accumulation, e.g., in association with in vitro neuronal differentiation. Therefore, by monitoring superoxide anion concentration in the culture medium after treatment with the chemotherapeutic agent, the authors were able to establish in a high throughput, non-invasive way the in vitro efficacy of 5-FU. This novel cell monitoring tool could be used for the accurate assessment of chemoresistance in cervical and other cancer cells, at least as far as its association with redox balance is concerned [33–35].

Within the same field of application and instead of measuring superoxide accumulation in cancer cells, Paivana et al. [36] opted for the direct assessment of the bioelectric properties of four different cancer cell lines (SK-N-SH, HEK293, HeLa and MCF-7) in response, once again, to 5-FU. Cancer cells were immobilized in calcium alginate matrix to mimic the natural tumor environment in vivo and cultured in different cell population densities (50,000 μL, 100,000 μL, and 200,000/100 μL). Bioelectric profiling was conducted by means of bioelectrical impedance-based measurements at three frequencies (1 KHz, 10 KHz, and 100 KHz). For impedance measurements, a voltage of 0.74 Vrms ± 50 mVrms was applied via the two terminals to the gold-coated electrodes. In this way, multi-dimensional mapping (cell line x population density x frequency) was achieved for the response of each cancer cell line against different 5-FU concentrations, in a rapid and entirely non-invasive way. It was demonstrated that bioimpedance measurements were highly correlated with standard cytotoxicity assays. This innovative bioimpedance profiling approach could enable the acquisition of a unique fingerprint for each cancer cell line response to a particular anticancer compound, therefore significantly accelerating the pace of chemotherapy drug screening.

The final contribution by Ibrahim et al. [37] is the one more closely related to the title of this editorial; namely, the integration of bioelectric sensors in the IoT networks and their role in the ongoing Digital or Industrial Revolution 4.0. In their report, the authors deal with the advanced yet quite an issue of protection against cyberattacks on remote health monitoring systems. In recent years, these systems have experienced almost incredible growth and popularity mainly due to their wide availability as fitness/daily life components of wearables and associated apps. On a more strictly medicinal level, IoT implantable therapeutic equipment and networks (availing over more than one hundred medical tools) are becoming standard issues of modern medical practice. One solution to counter cyberattacks, including tampering, sniffing, and unauthorized access is the construction of attack graphs as a technique to determine risks and vulnerabilities within interoperable systems and to identify possible attack paths. For this purpose, the authors used the pacemaker automatic remote monitoring system (PARMS) as a model for developing tailor-made attack graphs. They illustrate life-threatening risks to patients presented by hacking into the pacemaker's system and the feasibility of protecting implantable medical devices (IMDs) [38] by carrying out security strategies completely on an external device called a shield. This is definitely a technological field with considerable growth perspectives.

In conclusion, bioelectric sensors are here to stay in spite of their relatively recent emergence in diagnostic technology and related business. Without a doubt, they constitute an internal part of the wearables industry, which will keep on expanding in the next years. Bioelectric profiling is also becoming a valuable tool for rapid toxicity assays and compound x cell type fingerprinting, e.g., in the area of food safety control [10]. Innovative bioelectric sensors are being continuously developed to meet dire and yet unpreceded diagnostic and analytical needs; a vivid, very recent example is the expedient development of a cell-based bioelectric sensor for the ultra-sensitive detection of the SARS-CoV-2 S1 spike protein antigen in just three minutes [39]. As a final comment, bioelectric sensors may evolve as a separate scientific field themselves, opening new perspectives for a deeper understanding of bioelectric phenomena and their exploitation for practical purposes. One of the many possibilities in this direction is demonstrated by the new scientific topic of non-chemical distant cell interaction (NCDCI), where new principles of biology are being currently discovered in parallel with the development of innovative bioelectric sensing tools [40].

References

1. Bera, T.K. Bioelectrical impedance methods for non-invasive health monitoring: A review. *J. Med. Eng.* **2014**, *2014*, 381251. [CrossRef] [PubMed]
2. Nascimento, L.M.S.; Bonfati, L.V.; Freitas, M.L.B.; Mendes Junior, J.J.A.; Siqueira, H.V.; Stevan, S.L., Jr. Sensors and Systems for Physical Rehabilitation and Health Monitoring—A Review. *Sensors* **2020**, *20*, 4063. [CrossRef] [PubMed]
3. Liu, J.; Liu, M.; Bai, Y.; Zhang, J.; Liu, H.; Zhu, W. Recent Progress in Flexible Wearable Sensors for Vital Sign Monitoring. *Sensors* **2020**, *20*, 4009. [CrossRef] [PubMed]
4. Brosel-Oliu, S.; Abramova, N.; Uria, N.; Bratov, A. Impedimetric transducers based on interdigitated electrode arrays for bacterial detection—A review. *Anal. Chim. Acta* **2019**, *1088*, 1–19. [CrossRef]
5. Kramer, P.A.; Ravi, S.; Chacko, B.; Johnson, M.S.; Darley-Usmar, V.M. A review of the mitochondrial and glycolytic metabolism in human platelets and leukocytes: Implications for their use as bioenergetic biomarkers. *Redox Biol.* **2014**, *2*, 206–210. [CrossRef]
6. Souders, C.L., 2nd; Liang, X.; Wang, X.; Ector, N.; Zhao, Y.H.; Martyniuk, C.J. High-throughput assessment of oxidative respiration in fish embryos: Advancing adverse outcome pathways for mitochondrial dysfunction. *Aquat. Toxicol.* **2018**, *199*, 162–173. [CrossRef]

7. Crowe, S.M.; Kintzios, S.; Kaltsas, G.; Palmer, C.S. A Bioelectronic System to Measure the Glycolytic Metabolism of Activated CD4+ T Cells. *Biosensors* **2019**, *9*, 10. [CrossRef]
8. Neves, M.I.; Moroni, L.; Barrias, C.C. Modulating Alginate Hydrogels for Improved Biological Performance as Cellular 3D Microenvironments. *Front. Bioeng. Biotechnol.* **2020**, *8*, 665. [CrossRef]
9. Ferentinos, K.P.; Yialouris, C.P.; Blouchos, P.; Moschopoulou, G.; Kintzios, S. Pesticide residue screening using a novel artificial neural network combined with a bioelectric cellular biosensor. *BioMed Res. Int.* **2013**, *2013*, 813519. [CrossRef]
10. Moschopoulou, G.; Dourou, A.-M.; Fidaki, A.; Kintzios, S. Assessment of pesticides cytoxicity by means of bioelectric profiling of mammalian cells. *Environ. Nanotechnol. Monitor. Manag.* **2017**, *8*, 254–260. [CrossRef]
11. Kokla, A.; Blouchos, P.; Livaniou, E.; Zikos, C.; Kakabakos, S.E.; Petrou, P.S.; Kintzios, S. Visualization of the membrane engineering concept: Evidence for the specific orientation of electroinserted antibodies and selective binding of target analytes. *J. Mol. Recognit.* **2013**, *26*, 627–632. [CrossRef] [PubMed]
12. Dewey, J.A.; Dickinson, B.C. Split T7 RNA polymerase biosensors to study multiprotein interaction dynamics. *Methods Enzymol.* **2020**, *641*, 413–432. [PubMed]
13. Wiechert, J.; Gätgens, C.; Wirtz, A.; Frunzke, J. Inducible expression systems based on xenogeneic silencing and counter-silencing and design of a metabolic toggle switch. *ACS Synth Biol* **2020**, in press.
14. Kintzios, S. Consumer Diagnostics. In *Portable Biosensors and Point-of-Care Systems*; Kintzios, S., Ed.; IET: London, UK, 2017; pp. 309–331.
15. Kanakaris, G.P.; Sotiropoulos, C.; Alexopoulos, L.G. Commercialized point-of-care technologies. In *Portable Biosensors and Point-of-Care Systems*; Kintzios, S., Ed.; IET: London, UK, 2017; pp. 256–330.
16. Guo, J.; Liu, D.; Yang, Z.; Weng, W.; Chan, E.W.C.; Zeng, Z.; Wong, K.-Y.; Ling, P.; Chen, S. A photoelectrochemical biosensor for rapid and ultrasensitive norovirus detection. *Bioelectrochemistry* **2020**, *136*, 107591. [CrossRef] [PubMed]
17. Campuzano, S.; Pedrero, M.; Gamella, M.; Serafín, V.; Yáñez-Sedeño, P.; Pingarrón, J.M. Beyond sensitive and selective electrochemical biosensors: Towards continuous, real-time, antibiofouling and calibration-free devices. *Sensors* **2020**, *20*, 3376. [CrossRef]
18. Rodrigues, D.; Barbosa, A.I.; Rebelo, R.; Kwon, I.K.; Reis, R.L.; Correlo, V.M. Skin-Integrated Wearable Systems and Implantable Biosensors: A Comprehensive Review. *Biosensors* **2020**, *10*, 79. [CrossRef]
19. Yáñez-Sedeño, P.; Campuzano, S.; Pingarrón, J.M. Screen-Printed Electrodes: Promising Paper and Wearable Transducers for (Bio) Sensing. *Biosensors* **2020**, *10*, 76. [CrossRef]
20. Elsamnah, F.; Bilgaiyan, A.; Affiq, M.; Shim, C.-H.; Ishidai, H.; Hattori, R. Reflectance-Based Organic Pulse Meter Sensor for Wireless Monitoring of Photoplethysmogram Signal. *Biosensors* **2019**, *9*, 87. [CrossRef]
21. Kiani, M.; Du, N.; Vogel, M.; Raff, J.; Hübner, U.; Skorupa, I.; Bürger, D.; Schulz, S.E.; Schmidt, O.G.; Schmidt, H. P-N Junction-Based Si Biochips with Ring Electrodes for Novel Biosensing Applications. *Biosensors* **2019**, *9*, 120. [CrossRef]
22. Janssen, J.; Lambeta, M.; White, P.; Byagowi, A. Carbon Nanotube-Based Electrochemical Biosensor for Label-Free Protein Detection. *Biosensors* **2019**, *9*, 144. [CrossRef]
23. Lagoumintzis, G.; Zagoriti, Z.; Jensen, M.S.; Argyrakos, T.; Koutsojannis, C.; Poulas, K. Wireless Direct Microampere Current in Wound Healing: Clinical and Immunohistological Data from Two Single Case Reports. *Biosensors* **2019**, *9*, 107. [CrossRef] [PubMed]
24. Zhang, M.-L.; Wu, H.-T.; Chen, W.-J.; Xu, Y.; Ye, Q.-Q.; Shen, J.-X.; Liu, J. Involvement of glutathione peroxidases in the occurrence and development of breast cancers. *J. Transl. Med.* **2020**, *18*, 247. [CrossRef] [PubMed]
25. Brassart-Pasco, S.; Brézillon, S.; Brassart, B.; Ramont, L.; Oudart, J.B.; Monboisse, J.C. Tumor microenvironment: Extracellular matrix alterations influence tumor progression. *Front. Oncol.* **2020**, *10*, 397. [CrossRef] [PubMed]
26. Gao, L.; Loveless, J.; Shay, C.; Teng, Y. Targeting ROS-mediated crosstalk between autophagy and apoptosis in cancer. *Adv. Exp. Med. Biol.* **2020**, *1260*, 1–12. [PubMed]
27. Mavrikou, S.; Tsekouras, V.; Karageorgou, M.-A.; Moschopoulou, G.; Kintzios, S. Detection of Superoxide Alterations Induced by 5-Fluorouracil on HeLa Cells with a Cell-Based Biosensor. *Biosensors* **2019**, *9*, 126. [CrossRef] [PubMed]

28. Noordhuis, P.; Holwerda, U.; van der Wilt, C.L.; Groeningen, C.; Smid, K.; Meijer, S.; Pinedo, H.; Peters, G. 5-Fluorouracil incorporation into RNA and DNA in relation to thymidylate synthase inhibition of human colorectal cancers. *Ann. Oncol.* **2004**, *15*, 1025–1032. [CrossRef]
29. Walko, C.M.; Lindley, C. Capecitabine: A review. *Clin. Ther.* **2005**, *27*, 23–44. [CrossRef]
30. Hwang, P.M.; Bunz, F.; Yu, J.; Rago, C.; Chan, T.A.; Murphy, M.P.; Kelso, G.F.; Smith, R.A.; Kinzler, K.W.; Vogelstein, B. Ferredoxin reductase affects p53-dependent, 5-fluorouracil-induced apoptosis in colorectal cancer cells. *Nat. Med.* **2001**, *7*, 1111–1117. [CrossRef]
31. Fan, C.; Chen, J.; Wang, Y.; Wong, Y.S.; Zhang, Y.; Zheng, W.; Cao, W.; Chen, T. Selenocystine potentiates cancer cell apoptosis induced by 5-fluorouracil by triggering reactive oxygen species-mediated DNA damage and inactivation of the ERK pathway. *Free Radic. Biol. Med.* **2013**, *65*, 305–316. [CrossRef]
32. Liu, M.P.; Liao, M.; Dai, C.; Chen, J.F.; Yang, C.J.; Liu, M.; Chen, Z.G.; Yao, M.C. *Sanguisorba officinalis* L. synergistically enhanced 5-fluorouracil cytotoxicity in colorectal cancer cells by promoting a reactive oxygen species-mediated, mitochondria-caspase-dependent apoptotic pathway. *Sci. Rep.* **2016**, *27*, 34245. [CrossRef]
33. Chen, J.; Solomides, C.; Parekh, H.; Simpkins, F.; Simpkins, H. Cisplatin resistance in human cervical, ovarian and lung cancer cells. Cancer Chemother. *Pharmacology* **2015**, *75*, 1217–1227.
34. Liu, Y.; Li, Q.; Zhou, L.; Xie, N.; Nice, E.C.; Zhang, H.; Huang, C.; Lei, Y. Cancer drug resistance: Redox resetting renders a way. *Oncotarget* **2016**, *7*, 42740. [CrossRef] [PubMed]
35. Luo, M.; Wicha, M.S. Targeting cancer stem cell redox metabolism to enhance therapy responses. *Semin. Radiat. Oncol.* **2019**, *29*, 42–54. [CrossRef] [PubMed]
36. Paivana, G.; Mavrikou, S.; Kaltsas, G.; Kintzios, S. Bioelectrical Analysis of Various Cancer Cell Types Immobilized in 3D Matrix and Cultured in 3D-Printed Well. *Biosensors* **2019**, *9*, 136. [CrossRef] [PubMed]
37. Ibrahim, M.; Alsheikh, A.; Matar, A. Attack Graph Modeling for Implantable Pacemaker. *Biosensors* **2020**, *10*, 14. [CrossRef] [PubMed]
38. Gollakota, S.; Hassanieh, H.; Ransford, B.; Katabi, D.; Fu, K. They can Hear Your Heartbeats: Non-Invasive Security for Implantable Medical Devices. In Proceedings of the ACM SIGCOMM Conference, Toronto, ON, Canada, 15–19 August 2011; pp. 2–13.
39. Mavrikou, S.; Moschopoulou, G.; Tsekouras, V.; Kintzios, S. Development of a Portable, Ultra-Rapid and Ultra-Sensitive Cell-Based Biosensor for the Direct Detection of the SARS-CoV-2 S1 Spike Protein Antigen. *Sensors* **2020**, *20*, 3121. [CrossRef]
40. Apostolou, S.; Kintzios, S. Cell-to-Cell Communication: Evidence of Near-Instantaneous Distant, Non-Chemical Communication between Neuronal (Human SK-N-SH Neuroblastoma) Cells by Using a Novel Bioelectric Biosensor. *J. Conscious. Studies* **2018**, *25*, 62–74.

Article

Reflectance-Based Organic Pulse Meter Sensor for Wireless Monitoring of Photoplethysmogram Signal

Fahed Elsamnah [1], Anubha Bilgaiyan [2], Muhamad Affiq [1], Chang-Hoon Shim [2], Hiroshi Ishidai [3] and Reiji Hattori [1,4,*]

1 Department of Applied Science for Electronics and Materials, Kyushu University, Fukuoka 816-8580, Japan
2 COI STREAM, Center for Organic Photonics and Electronics Research (OPERA), Kyushu University, Fukuoka 819-0395, Japan
3 Konica Minolta, Inc., Ishikawa-cho, Hachioji 192-8505, Japan
4 Global Innovation Center (GIC), Kyushu University, Fukuoka 816-8580, Japan
* Correspondence: hattori@gic.kyushu-u.ac.jp; Tel.: +81-92-583-7887

Received: 19 June 2019; Accepted: 3 July 2019; Published: 10 July 2019

Abstract: This paper compares the structural design of two organic biosensors that minimize power consumption in wireless photoplethysmogram (PPG) waveform monitoring. Both devices were fabricated on the same substrate with a red organic light-emitting diode (OLED) and an organic photodiode (OPD). Both were designed with a circular OLED at the center of the device surrounded by OPD. One device had an OLED area of 0.06 cm^2, while the other device had half the area. The gap distance between the OLED and OPD was 1.65 mm for the first device and 2 mm for the second. Both devices had an OPD area of 0.16 cm^2. We compared the power consumption and signal-to-noise ratio (SNR) of both devices and evaluated the PPG signal, which was successfully collected from a fingertip. The reflectance-based organic pulse meter operated successfully and at a low power consumption of 8 μW at 18 dB SNR. The device sent the PPG waveforms, via Bluetooth low energy (BLE), to a PC host at a maximum rate of 256 kbps data throughput. In the end, the proposed reflectance-based organic pulse meter reduced power consumption and improved long-term PPG wireless monitoring.

Keywords: organic optoelectronic device; pulse meter; biosensor; Bluetooth low energy (BLE); photoplethysmogram (PPG)

1. Introduction

A pulse meter is a device used to measure the rate of rhythmic contraction and expansion of an artery at each beat of the heart based on the photoplethysmogram (PPG) principle. It has received enormous attention over the past decade, primarily from the healthcare industry, due to its continuous, real-time, and noninvasive monitoring, which provides the information necessary to determine an individual's health status and even provide a preliminary medical diagnosis [1–3]. Pulse meters rely on the PPG principle, which necessitates a light source and a light detector. The light is transmitted through tissue and reflects onto the light detector, as shown in Figure 1. When the heart beats, the blood volume of the arteries changes accordingly and causes variable light absorption, allowing changes in reflected light to be detected as a PPG signal. The detected PPG signal comprises an alternating (AC) component, due to the variable absorption of the pulsatile arterial blood, and a steady-state (DC) component, from the veins, capillaries, tissues, bones, and other non-pulsatile components, as shown in Figure 2 [4]. The AC component is the outcome of light absorption by the arteries, while the DC component is the outcome of light absorption by body tissues and veins. Therefore, the pulsatile effect occurs only in the arteries, not in the veins or other non-pulsatile components. There are two approaches that can be used to obtain a PPG signal from a biosensor pulse meter: reflection and transmission. The reflection method was utilized in this work because of the freedom of use. The device could be easily worn or attached to different parts of the human body. The transmission method involves tissue transillumination and required that a light source and a detector be placed opposite each other. Consequently, the transmission method could only be used on external body parts such as fingertips and ear lobes.

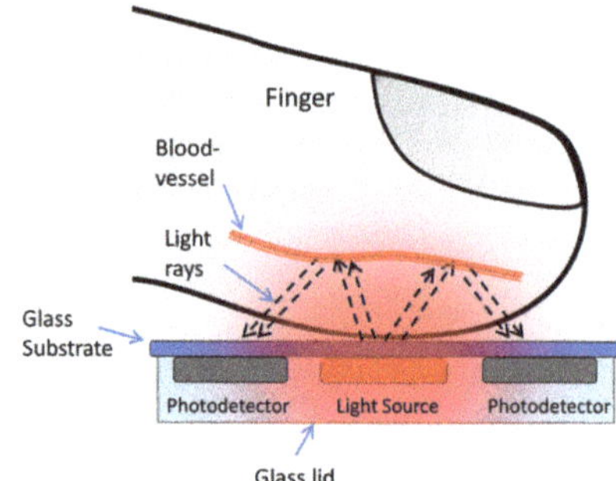

Figure 1. Acquisition of photoplethysmogram (PPG) signal from the reflection method.

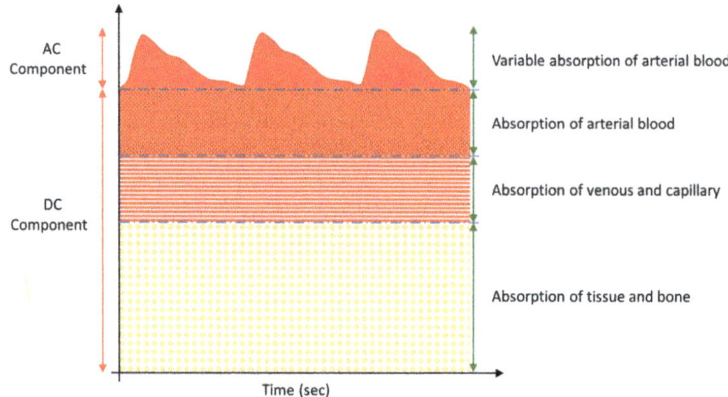

Figure 2. Schematic of light absorption in body tissues.

In recent years, organic pulse oximeters have received significant attention from researchers due to the many advantages of organic optoelectronic devices, including their relative low cost, simple fabrication and their ability to be fabricated on flexible substrates, for comfortable wearable medical devices. Furthermore, large organic photodiodes (OPDs) can be easily fabricated, compared to the restricted size of generic silicon-based photodiodes (PD). This has made organic light-emitting diodes (OLEDs) and OPDs preferable for use in wearable pulse oximeters [5–7]. In the literature, there were several proposed OLED and OPD designs that aimed to improve power consumption and signal quality. To improve the longevity of the batteries in inorganic reflective pulse oximeters, the authors of [8] proposed an annular PD ring design with a light-emitting diode (LED) located at the center. A rectangular OPD device and a device with two separated square OLEDs were proposed in [9]. Meanwhile, the authors in [10] proposed a design with a circular OPD in the center of a half-ring of red polymer light-emitting diodes (PLEDs) and a half-ring of green PLEDs. The authors of [11] conducted optical simulations to test the power consumption of their designs involving a ring of OPDs surrounding a circular OLED. Various other researchers have attempted to develop a wireless pulse meter and to solve the problems associated with it, such as signal quality and power consumption. In [12], the study proposed a compact portable module composed of an array of photodetectors that could be distributed radially around LEDs and the PPG signal sent via a Zigbee protocol wireless module. The chip consumed 38 mA to transmit the data and 37 mA to receive the data. The red LEDs consumed about 38 mW and the IR LEDs about 26 mW. The authors of [13] proposed a wireless heart rate (HR) and peripheral oxygen saturation (SpO_2) monitoring system that could be connected to a local wireless network via Wi-Fi technology and the information was transmitted in real time to a webpage for remote monitoring. The current consumption of that wireless microcontroller unit (MCU) chip was 229 mA for transmission (TX) traffic and 59 mA for reception (RX) traffic. Other researchers proposed wireless pulse oximeters but did not mention the power consumption of the proposed device, such as [14], who proposed a wireless ring-type pulse oximeter with multiple detectors for sending the signal to the host system via Bluetooth. In [15], the authors presented a PPG wireless monitoring device embedded in a hat and glove that could send the signal via Bluetooth. However, the wireless PPG signal quality was not adequately addressed in the previous research and prototypes. Moreover, power consumption is a top priority in the development of wireless pulse meters because they are battery operated. Although the previous works on pulse oximeters required two light sources, which consumed double the power of one light source. The previously proposed wireless devices remained impractical for long-term use, although the pulse oximeters required two light sources, which consumed double the power of one light source. Therefore, miniature, portable, wearable pulse meters that are able to be monitored wirelessly will provide more freedom and comfort and will be more compatible with conventional devices, which will lead to simplified health monitoring [16]. In terms of organic optoelectronic devices, the material structure, dimension design, and the characteristics of the OPDs and OLEDs in this paper were different to those in previous works. Here, we propose different material structures and dimensions of OLED/OPD devices based on our previous works [17–19] as part of our continued attempts to improve the power consumption and signal quality of organic pulse oximeter.

In this work, we compared two different organic pulse meter designs using OLEDs and OPDs. We evaluated their performances in terms of the power consumption of the wireless pulse meter. This paper highlighted the importance of designing OPD and OLED structures, guided by optical simulation, to enhanced signal quality and minimize the power consumption for monitoring the PPG waveforms via Bluetooth low energy (BLE).

2. Materials and Methods

Two devices, Device-1 and Device-2, were fabricated with different OLED areas. Both devices had a circular OLED at the center and were surrounded by a ring of OPDs. Device-1 had an OLED area of 0.06 cm^2 while Device-2 had an OLED area of 0.03 cm^2. The gap distance between the OLED and OPD was 1.65 mm and 2 mm for Device-1 and Device-2, respectively. Both devices had the same OPD area of 0.16 cm^2. The OPDs were ring-shaped and surrounded the OLED to effectively collect the reflected photons from the skin.

2.1. Optical Simulation

To simulate the whole system optically, a simplified finger model was designed and configured, followed by the OLED and OPD device models. Next, all models were simulated optically using LightTools software (Synopsys, Inc., Mountain View, CA, USA), which uses the ray-tracing and Monte Carlo methods. Figure 3 illustrates the distribution of the light rays from the red OLED into the human finger and shows the simplified four-layer structure of a finger model. The red OLED was adopted due to its advantages over other OLEDs colors including the green OLED and near-infrared (NIR) OLED. In the case of green OLED, the green light has a shorter wavelength, which leads to decreasing the penetration depth of the light in the human body and getting absorbed. In the case of the NIR OLED, the NIR region of the spectrum has a longer wavelength, which leads to increasing the penetration depth of the light in the human body. However, the EQE in this region is significantly lower than the OLED in the visible region. Moreover, fabricating NIR OLED/OPD is difficult and still under research and development [20,21]. In the model, the arterial blood vessels were assumed to be included in the skin and the subcutaneous adipose tissue for simplicity. Device-1 and Device-2 were attached to the simplified finger model as shown in Figure 3A,B. The optical parameters were approximated using the literature [22–25], and are shown in Table 1 where the refractive index of the material describes how fast light propagates through the material. The Henyey–Greenstein function was used to approximate the angular scattering dependence of single scattering events in biological tissues.

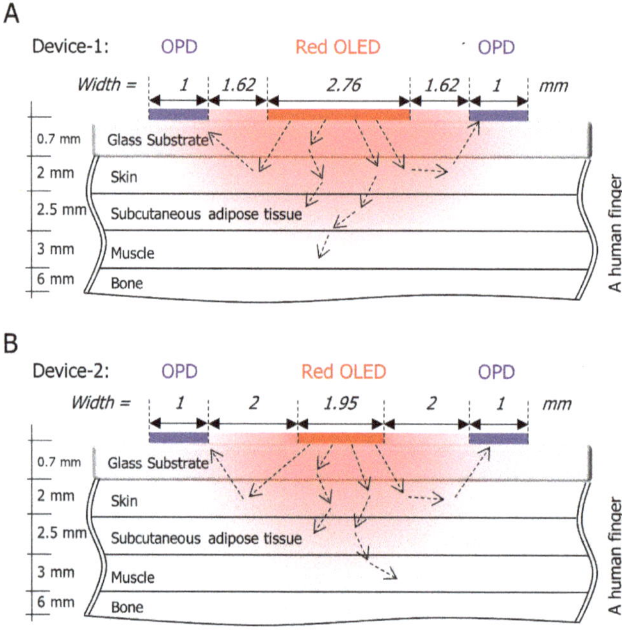

Figure 3. A simplified finger schematic with the organic light-emitting diode (OLED) as the light source and the organic photodiodes (OPDs) as the surface receiver for (**A**) Device-1 and (**B**) Device-2.

Table 1. Summary of the approximated optical parameters of a human finger.

Tissue	Wave-length (nm)	Index of Refraction (n)	Henyey–Greenstein (g)	Absorption Coefficient (Ua) in mm^{-1}	Scatter Coefficient (Us) in mm^{-1}	Thickness (mm)
Human Skin	625	1.55	0.81	0.27	18.7	2
Subcutaneous Fat	625	1.44	0.9	1.14	12.8	2.5
Muscle	625	1.37	0.9	0.56	64.7	3
Bone	625	1.37	0.9	0.04	19.5	6

The dimensions of the OLED and OPD design structure for both devices are illustrated in Figure 4A,B, which demonstrates that the area of the OLED in Device-1 was designed to be double the size of that in Device-2 in order to evaluate the effect of increasing the light source and decreasing the gap distance between the OLED and the OPDs. The radiant power of the light source was assumed, based on our previous red OLED devices (625 nm), to be 2.8 µW and 1.4 µW for Device-1 and Device-2, respectively. The simulation traced one million rays in each device. The simulation results, shown in Figure 4C,D, show the irradiance on the OPD area, where the maximum estimated irradiance for Device-1 was 9.8×10^{-10} W/mm^2 and for Device-2 was 3.7×10^{-10} W/mm^2. While the simulation presented an unequal irradiance distribution due to the spline effect of the ray simulation, we assumed the maximum irradiance was distributed equally. Therefore, the total received power of Device-1 and Device-2 was 15.6×10^{-9} W and 5.9×10^{-9} W, respectively. To put that into perspective, 0.55% of light rays reflected on the OPD of Device-1, while Device-2 received 0.42% of the light rays on the OPD. In these simulation results, when the gap between the OLED and the OPDs was decreased from 2 mm in Device-2 to 1.62 mm in Device-1, the total power increased to more than double, which means that the amplitude of the PPG in Device-1 was expected to be bigger than the amplitude in Device-2.

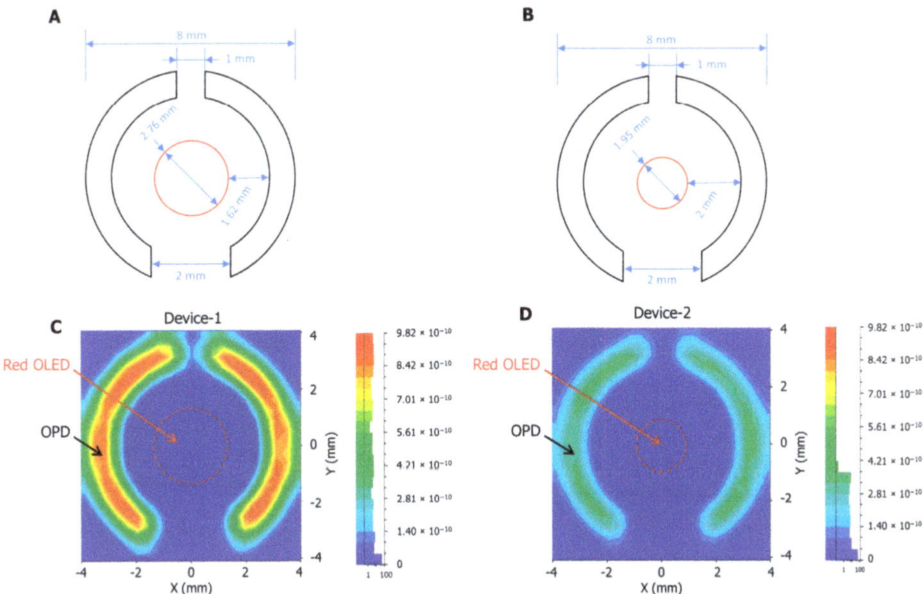

Figure 4. The dimensions of the OLED and OPD design structure in (**A**) Device-1 and (**B**) Device-2. The optical simulation results at the receiver's surface in Watt per mm^2 for (**C**) Device-1 and (**D**) Device-2.

2.2. The Organic Optoelectronic Device

Device-1 and Device-2 were fabricated with two different OLED sizes and a similar OPD area. As a monolithic device, the OLED and the OPD were deposited onto the same 0.7-mm-thick glass substrate. For simplicity of the structure of the device and future fabrication, identical red OLED and OPD structures, shown in Figure 5A, were fabricated on the same substrate. The organic material structure of the OLEDs and OPDs were the same for Device-1 and Device-2, as shown in Figure 5B.

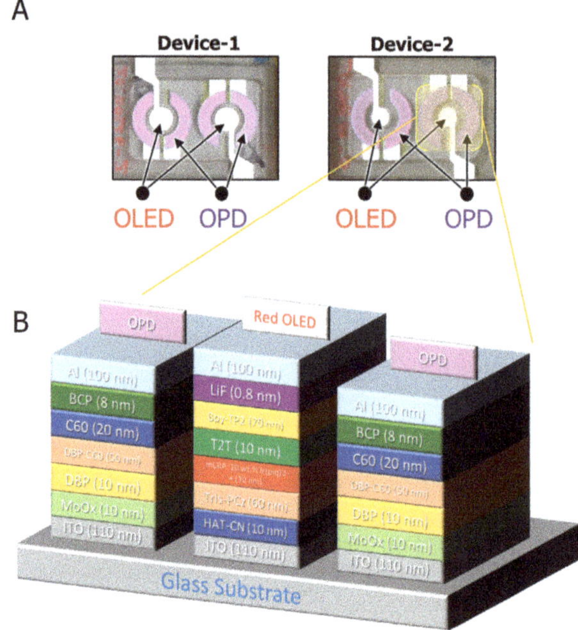

Figure 5. (**A**) The fabricated organic biosensor pulse meter for Device-1 and Device-2. (**B**) A cross-section of the organic material structure of the OLED and OPD for both devices.

The fabrication process of the organic pulse meter sensor started with the preparation of glass substrates that were coated with indium-tin-oxide (ITO) and then loaded into an evaporator chamber to deposit organic metal layers with the proper shadow mask patterns for the OLED and followed by the OPD. The shadow masks were designed in a way to prevent the issue of misalignment in the multi-mask processing of monolithic device fabrication. Then, the samples were encapsulated without exposure to air in a glove box, which was connected to the deposition chamber by glass lid using UV curable epoxy resin [18]. Thereafter, the samples were used for evaluating the performance of the pulse meter sensor. Figure 5B illustrates the structure of the OLED and OPD used in Device-1 and Device-2. The OLED device structure consisted of ITO (110 nm), a transparent anode that allowed the generated light to be emitted from the glass panel, HAT-CN (10 nm) as the hole injection layer, Tris-PCz (60 nm) as the hole transport layer, mCBP:10 wt% Ir(piq)$_3$ + (30 nm) as the emissive layer where the holes and electrons recombined to emit light, T2T (10 nm) as the electron blocking layer, Bpy-TP2 (70 nm) as the electron transport layer, LiF (0.8 nm) as the electron injection layer, and Al (100 nm) as the cathode for the OLED. The OPD structure consisted of ITO (110 nm), with MoO$_x$ (10 nm) as the hole injection layer, DBP (10 nm), DBP:C60 (50 nm), and C60 (20 nm) combined in the device active layer, BCP (8 nm) as the electron transport layer, and Al (100 nm) as the cathode.

The characteristics of the OLED and OPD devices were essential to the signal quality of the biosensor pulse meter. Organic LEDs have many advantages over LEDs, despite the fact that they have a shorter lifetime. OLEDs are relatively easy to produce, flexible, foldable, transparent, and capable of fast switching. Moreover, their cost is lower and they consume less power. The emission spectra of the OLED for Device-1 and Device-2, where the maximum intensity of light was 625 nm, are shown in Figure 6A. The slight difference in the characteristics of both devices was due to thickness variation because the devices were not fabricated at the same time. Figure 6B presents the OPD's external quantum efficiency (EQE) at zero-bias condition, where the maximum EQE was about 37% at 625 nm, which sufficiently overlapped with the peak EL of the red OLED wavelength.

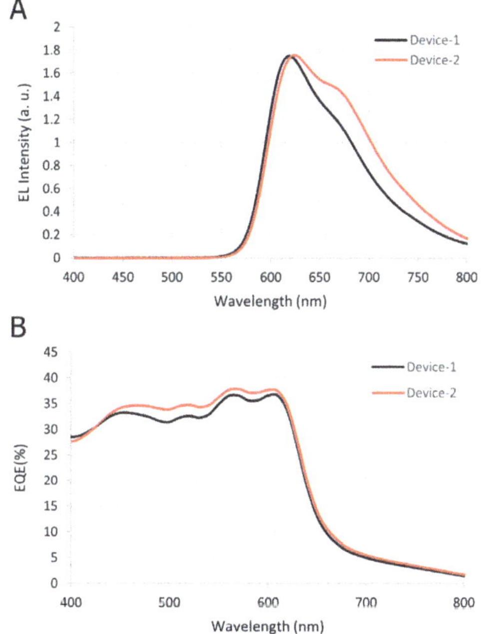

Figure 6. (**A**) The OLED's electroluminescence spectrum with respect to wavelength (nm). (**B**) The OPD's external quantum efficiency (EQE) with respect to the wavelength.

2.3. The Device Structure

The structure of the proposed biosensor pulse meter for wireless monitoring was composed of three parts: an organic optoelectronic device, a device holder, and a PCB that included a driver circuit and a wireless microcontroller unit (MCU) module, as illustrated in Figure 7.

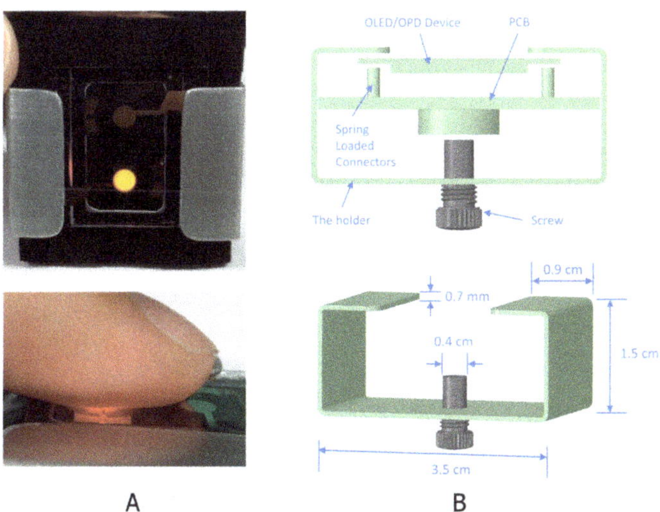

Figure 7. (**A**) The proposed portable pulse meter prototype for wireless monitoring via Bluetooth low energy (BLE). (**B**) The device holder, that used to fix the parts together, and its dimensions.

In the driver circuit, the small photocurrent generated by the OPD needed a transimpedance amplifier (TIA) to convert the OPD current to a voltage signal. Next, the PPG signal required a filtration process to eliminate the high-frequency noise and the DC component from the signal. The last stage was to amplify the AC component of the PPG signal and send it through an analog-to-digital converter (ADC) process. Then, the PPG signal was transmitted via serial communication block (SCB) or Bluetooth low energy (BLE). The analog circuit employed in this work was previously described, see [17] for more details. Figure 7A presents the portable pulse meter prototype that was proposed for wireless monitoring of the PPG signal, and Figure 7B illustrates the dimensions of our novel tool for fixing the biosensor pulse meter and the PCB, including the driver circuit and wireless MCU together with aluminum shield for further protection. The wireless MCU module (CYBLE-214015-01, Cypress Semiconductor, San Jose, CA, USA) implemented in this work is a certified and qualified module supporting BLE wireless communication. It uses a BLE technology in which the battery life is enhanced by keeping the radio activity short and allowing the device to reside in standby or power-down mode during most of its operating time to consume less power.

3. Results and Discussion

3.1. Comparative Results for Device-1 and Device-2

The two pulse meters were evaluated in vivo on a healthy male subject, on his index finger. The PPG signal was collected from Device-1 and Device-2 sequentially, within a specific time period, while he was sitting on a chair. The data were recorded at the sampling frequency of 500 SPS, 8-bit resolution from the SCB and used to evaluate both devices. Figure 8 shows that both devices were reliable and obtained a clear and stable PPG signal. With Device-1, Figure 8A, the peak-to-peak amplitude (V_{pp}) of the PPG signal was about 100 mV, while in Figure 8B, the PPG amplitude of Device-2 was less than 45 mV with a constant voltage of 5 V of the OLED. Consequently, the power consumption of each device was different. Device-1 consumed about 1.6 mW and Device-2 about 0.6 mW. Therefore, in order to unify the power consumption of both devices, we supplied a constant current. We compared the amplitude and signal-to-noise ratio (SNR) values of both devices for each PPG signal at different OLED's driving currents, from 1.2 μA to 93.6 μA, as shown in Figure 9. The constant current was supplied using the MCU's analog current source. Table 2 summarizes the PPG signal characteristics of Device-1 and Device-2. The method for quantifying the PPG signal quality using the SNR measurement was previously described in [17].

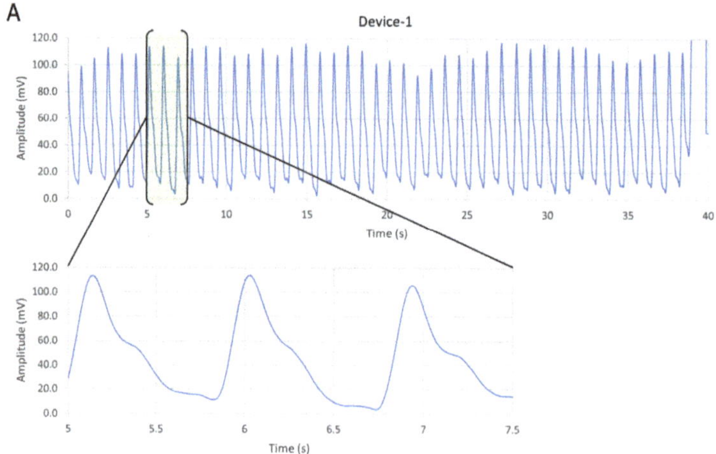

Figure 8. *Cont.*

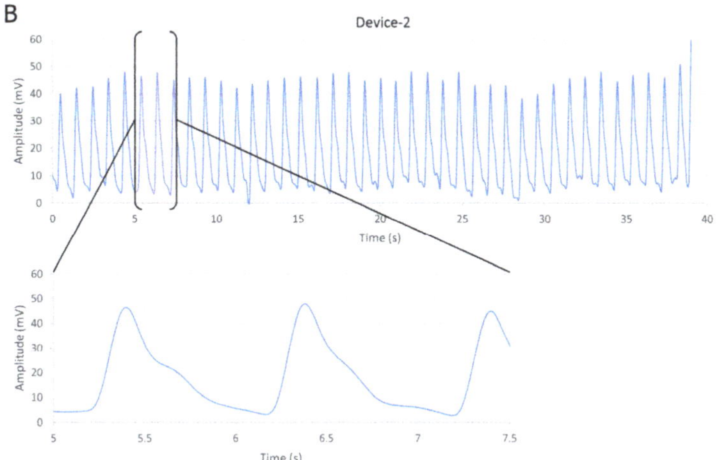

Figure 8. Obtaining the PPG signal at a constant voltage source from (**A**) Device-1 and (**B**) Device-2.

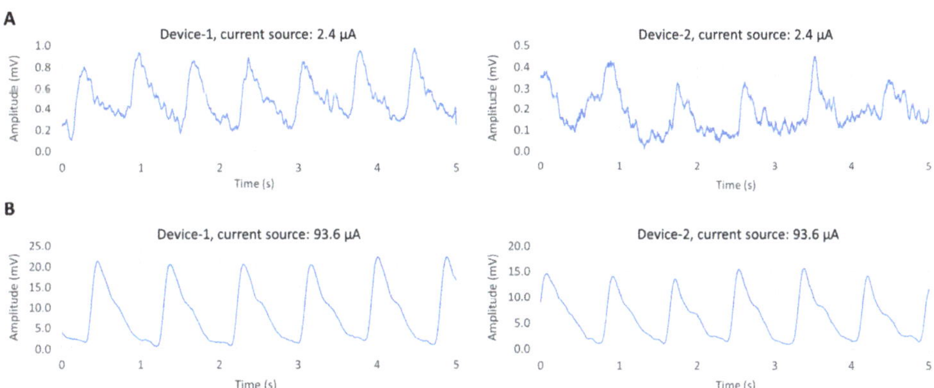

Figure 9. Comparison between the PPG signals from both devices at two different current sources for OLED (**A**) at 2.4 μA and (**B**) at 93.6 μA.

Table 2. Summary of the PPG signal quality for Device-1 and Device-2.

Device No.	Average V_{pp} (mV)	SNR (dB)	Current Source (μA)
Device-1	20	45	93.6
Device-2	13	46	93.6
Device-1	0.7	18	2.4
Device-2	0.3	8	2.4

Figure 10 compares the SNR values from Device-1 and Device 2 of different gap distance at different current consumptions. It shows that decreasing the distance of the gap between the OLED and the OPD, in Device-1, leads to increasing in the DC noise on the OPD, as was indicated by the SNR results with respect to the OLED current. The SNR of Device-2 at a higher OLED current of 93.6 μA was 45 dB, whereas it was 46 dB in Device-1 at the same current source even though its PPG signal amplitude was smaller than Device-2. On the other hand, reducing the gap between the OLED and the OPD resulted in a higher SNR of the PPG signal at a lower OLED current. As shown in Table 2, the PPG signal amplitude increased from 0.3 mV at 8 dB in Device-2 to 0.7 mV at 18 dB in Device-1, where the gap distances between the OLED and the OPDs were 1.65 mm and 2 mm for

Device-1 and Device-2, respectively. It is worth mentioning that Device-1 produced an acceptable PPG signal of about 18 dB with ultra-low power consumption, as low as 8 μW. Consequently, from these results, Device-1 demonstrated a significantly advantageous SNR at a low current supply compared to Device-2 where the short gap between the OLED and the OPD increased the chance of obtaining more reflected light. That means the shorter gap between the OLED and the OPD devices will result in improving the SNR at the low power consumption, and the larger gap will result in reducing the SNR.

Although the gap distance between the OLED and the OPD is a top priority while designing an organic pulse meter, another two aspects that should be taken into consideration are the OLED's area, and OPD's area. For the OLED's area, if we reduced the OLED's area in order to reduce the power consumption, the OLED's lifetime will also be reduced because the current density will be increased on a small area. On the other hand, if the OLED's area increased, then the area in the center of the OLED will be impractical and will not contribute to increasing the reflected light from the human body. For the OPD's area, it should be designed in a way to surround the OLED sufficiently in order to collect more reflected light. However, increasing the OPD's area too much will result in increasing the DC noise. Therefore, the area of the OPD should be consistent with OLED design, which can be achieved by optical simulation.

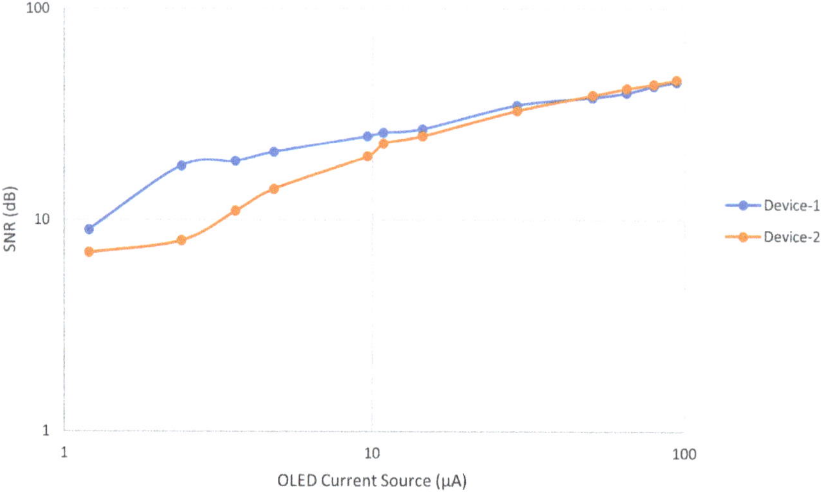

Figure 10. Comparison of the signal-to-noise ratio (SNR) of the PPG signals from Device-1 and Device-2 at different OLED driving currents from 1.2 μA to 93.6 μA.

Table 3 compares the results between our device and other devices, despite the fact that the previous PPG signal monitoring systems are different than our current proposed device in terms of device characteristics, flexibility and signal quality. The proposed device consumes the lowest power for light source among them.

Table 3. Comparison table between our device and other devices.

	This Work	Reference [11]	Reference [17]	Reference [10]	Reference [9]
OLED Type	Red OLED	Red OLED	Red OLED	Red PLED	Red OLED
Device Flexibility	Rigid	Flexible	Rigid	Flexible	Rigid
Voltage Supply (V)	3.3	3.3	5	5	9
OLED Driving Current (μA)	2.4	21	20	1000	20000
OLED Area (mm^2)	6	0.5	3	N.C.	4
Power Consumption (μW)	8	24	100	N.C.	N.C.
PPG Signal-to-Noise Ratio (dB)	18	N.C.	45	N.C.	N.C.

3.2. Results of BLE PPG Signal from Device-1

This section presents the results of the wireless PPG waveforms from Device-1, which was selected as the more effective device, in terms of power consumption, for the portable pulse meter. The PPG signal was obtained from the index finger of the subject while he was resting in a chair during a specific period of time. The data were obtained wirelessly from the portable organic pulse meter using BLE technology. The PPG signal was sent to a PC host using a universal serial bus (USB) dongle (PRoC BLE CYBL10162-56LQXI, Cypress Semiconductor, San Jose, CA, USA) with 500 SPS of ADC and 8-bit resolution. The recorded length of the PPG signal on the receiving PC host was constrained by the sampling rate of the ADC in the chip. Although the BLE chip had an on-air data rate of 1000 kbps, the maximum throughput data rate between the chip and our receiving PC host was 256 kbps at a minimum connection interval of 7.5 ms. The maximum throughput was measured using the PSoC Creator program from [26] on the BLE module and displayed on terminal window software (Tera Term, Ver. 4.93).

Device-1 successfully showed a very clear PPG signal on the receiving PC host. Figure 11A shows the wireless PPG waveform on our C# program where the data was obtained from the USB dongle that supported BLE. The role of the USB dongle is to convert the Bluetooth signal to a COM port serial signal. Therefore, the dongle was needed in the case that the PC did not have a built-in Bluetooth or did not support the BLE. PPG waveform was recorded from the program as a comma-separated values (CSV) file and is presented in Figure 11B. It is noteworthy to mention that the PPG signals encounter several factors that can influence the signal's quality such as the amount of pressure from the human body against OLED/OPD device substrate. The device substrate should be coupled with the skin in order to receive the reflected light on the OPD. However, more pressure on the skin results in more pressure on the blood vessels and that causes weakness of the PPG signal's amplitude. Therefore, moderate pressure from the human body should be applied to the pulse meter's substrate.

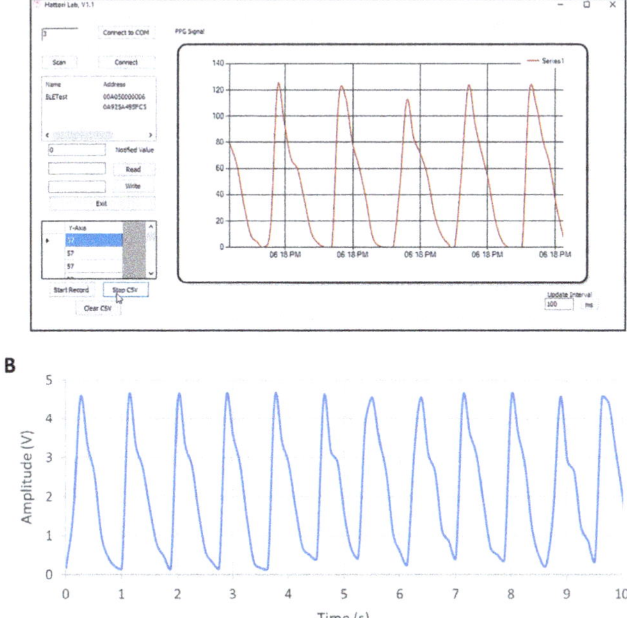

Figure 11. (**A**) The C# program shows PPG signal obtained from a universal serial bus (USB) dongle that was connected to the portable pulse meter via BLE. (**B**) PPG waveform after being recorded from the program as a comma-separated values (CSV) file.

4. Conclusions

This paper compared two structural designs of organic pulse meters, Device-1 and Device-2, in order to minimize the power consumption in wireless monitoring of PPG waveforms. We discussed the optical simulation results of both devices. The proposed designs were simulated using ray-tracing and Monte Carlo methods to evaluate the effect of increasing the light source area and decreasing the gap distance between the OLED and OPD. Based on the simulation, the total optical power of Device-1 was more than double to that of Device-2, where the gap distance between the OLED and OPD was 2 mm and 1.62 mm for Device-1 and Device-2, respectively. The simulation results were verified by fabricating two pulse meter structures with different circular OLED areas at the center of the device but surrounded by same ring-shaped OPD area. The performance of the proposed devices was tested in vivo on a healthy individual. In the experimental results, reducing the gap between the OLED and the OPDs resulted in a higher SNR of the PPG signal at a low OLED power source, and a slightly lower SNR at a high OLED power source due to the DC noise. The biosensor pulse meter showed promising results with ultra-low power consumption, 8 μW at 18 dB SNR, and demonstrated its ability to measure a clear PPG signal up to 46 dB SNR at constant current of 93.6 μA. The proposed reflectance-based organic pulse meter sensor was used to wirelessly monitor the PPG signals, and its compatible characteristics were successfully demonstrated. Clear PPG waveforms were obtained from the portable pulse meter via BLE at 500 SPS and 8-bit resolution on the receiving PC host. The maximum throughput data rate between the chip and the PC host was 256 kbps at the minimum connection interval of 7.5 ms. Our proposed device was capable of producing a clear PPG signal and operating on ultra-low power, which is essential for long-term wireless PPG signal monitoring. In future work, the organic optoelectronic device, OPD/OLED, will be fabricated onto a flexible substrate in order to add flexibility to the pulse meter as well as comfortability of use as a wearable medical device.

Author Contributions: Supervision and validation, R.H.; conceptualization, all authors; formal analysis, F.E. and M.A.; methodology, A.B., H.I., C.-H.S., and F.E.; software, F.E.; writing—original draft preparation F.E.; writing—review and editing, R.H. and F.E. All authors approved the final manuscript.

Funding: This research was funded by the Center of Innovation Program (COI STREAM) from the Japan Science and Technology Agency (JST).

Acknowledgments: Support from Kyushu University and the Ministry of Education, Culture, Sports, Science, and Technology (MEXT), Japan is highly appreciated.

Conflicts of Interest: The authors declare no conflict of interest. The funder had no role in the design of the study, the collection, analysis, or interpretation of the data, the writing of the manuscript, or the decision to publish the results.

References

1. Webster, J.G. *Design of Pulse Oximeters*; CRC Press: Boca Raton, FL, USA, 1997.
2. Jubran, A. Pulse oximetry. In *Applied Physiology in Intensive Care Medicine 1*, 3rd ed.; Physiological Notes—Technical Notes—Seminal Studies in Intensive Care; Springer-Verlag: Berlin/Heidelberg, Germany, 2012; pp. 51–54.
3. Guk, K.; Han, G.; Lim, J.; Jeong, K.; Kang, T. Evolution of Wearable Devices with Real-Time Disease Monitoring for Personalized Healthcare. *Nanomaterials* **2019**, *9*, 813. [CrossRef] [PubMed]
4. de Kock, J.P.; Tarassenko, L. Pulse oximetry: theoretical and experimental models. *Med. Biol. Eng. Comput.* **1993**, *31*, 291–300. [CrossRef] [PubMed]
5. Chen, C.T. Evolution of red organic light-emitting diodes: Materials and devices. *Chem. Mater.* **2004**, *16*, 4389–4400. [CrossRef]
6. Thejo Kalyani, N.; Dhoble, S.J. Organic light emitting diodes: Energy saving lighting technology—A review. *Renew. Sustain. Energy Rev.* **2012**, *16*, 2696–2723. [CrossRef]
7. Geffroy, B.; le Roy, P.; Prat, C. Organic light-emitting diode (OLED) technology: Materials, devices and display technologies. *Polym. Int.* **2006**, *55*, 572–582. [CrossRef]

8. Mendelson, Y.; Pujary, C. Measurement site and photodetector size considerations in optimizing power consumption of a wearable reflectance pulse oximeter. In Proceedings of the 25th Annual International Conference of the IEEE Engineering in Medicine and Biology Society, Cancun, Mexico, 17–21 September 2003; pp. 3016–3019.
9. Lochner, C.M.; Khan, Y.; Pierre, A.; Arias, A.C. All-organic optoelectronic sensor for pulse oximetry. *Nat. Commun.* **2014**, *5*, 1–7. [CrossRef] [PubMed]
10. Tachibana, Y.; Kaltenbrunner, M.; Koizumi, M.; Matsuhisa, N.; Yukita, W.; Zalar, P.; Jinno, H.; Someya, T.; Kitanosako, H.; Yokota, T. Ultraflexible organic photonic skin. *Sci. Adv.* **2016**, *2*, e1501856.
11. Kim, H.; Lee, H.; Yoo, S.; Kim, M.; Kim, E.; Lee, J.; Yoo, H.-J.; Lee, Y. Toward all-day wearable health monitoring: An ultralow-power, reflective organic pulse oximetry sensing patch. *Sci. Adv.* **2018**, *4*, eaas9530.
12. Li, K.; Warren, S. A wireless reflectance pulse oximeter with digital baseline control for unfiltered photoplethysmograms. *IEEE Trans. Biomed. Circuits Syst.* **2012**, *6*, 269–278. [CrossRef] [PubMed]
13. Ayance, T.; Trevi, C.G. Wireless heart rate and oxygen saturation monitor Wireless Heart Rate and Oxygen Saturation Monitor. *AIP Conf. Proc.* **2019**, *2090*, 1–5.
14. Huang, C.; Chan, M.; Chen, C.; Lin, B. Novel Wearable and Wireless Ring-Type Pulse Oximeter with Multi-Detectors. *Sensors* **2014**, *14*, 17586–17599. [CrossRef] [PubMed]
15. Spigulis, J.; Erts, R.; Nikiforovs, V.; Kviesis-kipge, E. Wearable wireless photoplethysmography sensors. In Proceedings of the Biophotonics: Photonic Solutions for Better Health Care, Strasbourg, France, 7–11 April 2008; Volume 6991, pp. 1–7.
16. Ha, M.; Lim, S.; Ko, H. Wearable and flexible sensors for user-interactive health-monitoring devices. *J. Mater. Chem. B* **2018**, *6*, 4043–4064. [CrossRef]
17. Elsamnah, F.; Bilgaiyan, A.; Affiq, M.; Shim, C.-H.; Ishidai, H.; Hattori, R. Comparative Design Study for Power Reduction in Organic Optoelectronic Pulse Meter Sensor. *Biosensors* **2019**, *9*, 48. [CrossRef] [PubMed]
18. Bilgaiyan, A.; Sugawara, R.; Elsamnah, F.; Shim, C.; Affiq, M.; Hattori, R. Optimizing performance of reflectance-based organic Photoplethysmogram (PPG) sensor. In Proceedings of the Organic and Hybrid Sensors and Bioelectronics XI, San Diego, CA, USA, 19–23 August 2018; Volume 10738, p. 1073808.
19. Elsamnah, F.; Hattori, R.; Shim, C.-H.; Bilgaiyan, A.; Sugawara, R.; Affiq, M. Reflectance-based Monolithic Organic Pulsemeter Device for Measuring Photoplethysmogram Signal. In Proceedings of the 2018 IEEE International Instrumentation and Measurement Technology Conference (I2MTC), Houston, TX, USA, 14–17 May 2018; pp. 1–5.
20. Sommer, J.R.; Farley, R.T.; Graham, K.R.; Yang, Y.; Reynolds, J.R.; Xue, J.; Schanze, K.S. Efficient Near-Infrared Polymer and Organic Light-Emitting Diodes Based on Electrophosphorescence from. *ACS Appl. Mater. Interfaces* **2009**, *1*, 274–278. [CrossRef] [PubMed]
21. Xue, J.; Li, C.; Xin, L.; Duan, L.; Qiao, J. High-efficiency and low efficiency roll-off near-infrared fluorescent OLEDs through triplet fusion. *Chem. Sci.* **2016**, *7*, 2888–2895. [CrossRef] [PubMed]
22. Jacques, S.L. Optical properties of biological tissues: a review. *Phys. Med. Biol.* **2013**, *58*, R37. [CrossRef] [PubMed]
23. Zamburlini, M.; Pejović-Milić, A.; Chettle, D.R.; Webber, C.E.; Gyorffy, J. In vivo study of an x-ray fluorescence system to detect bone strontium non-invasively. *Phys. Med. Biol.* **2007**, *52*, 2107–2122. [CrossRef] [PubMed]
24. Akkus, O.; Uzunlulu, M.; Kizilgul, M. Evaluation of Skin and Subcutaneous Adipose Tissue Thickness for Optimal Insulin Injection. *J. Diabetes Metab.* **2012**, *3*. [CrossRef]
25. Drahansky, M.; Kanich, O.; Brezinová, E.; Shinoda, K. Experiments with Optical Properties of Skin on Fingers. *Int. J. Opt. Appl.* **2016**, *6*, 37–46.
26. Luthra, G. PROJECT #024: BLE Throughput—Pushing the Limits. Available online: http://www.cypress.com/blog/100-projects-100-days/project-024-ble-throughput-pushing-limits (accessed on 10 June 2019).

Article

P-N Junction-Based Si Biochips with Ring Electrodes for Novel Biosensing Applications

Mahdi Kiani [1,*], Nan Du [1,*], Manja Vogel [2], Johannes Raff [2], Uwe Hübner [3], Ilona Skorupa [2], Danilo Bürger [1], Stefan E. Schulz [1,4], Oliver G. Schmidt [5] and Heidemarie Schmidt [1,3,*]

1 Department Back-End of Line, Fraunhofer Institute for Electronic Nano Systems, Technologie-Campus 3, 09126 Chemnitz, Germany
2 Helmholtz-Zentrum Dresden-Rossendorf, Bautzner Landstraße 400, 01328 Dresden, Germany
3 Leibniz Institute of Photonic Technology, Albert-Einstein-Str. 9, 07745 Jena, Germany
4 Center for Microtechnologies, Chemnitz University of Technology, Reichenhainer Str. 70, 09126 Chemnitz, Germany
5 Institute for Integrative Nanosciences IFW Dresden, Helmholtzstr. 20, 01069 Dresden, Germany
* Correspondence: mahdi.kiani@enas.fraunhofer.de (M.K.); nan.du@enas.fraunhofer.de (N.D.); heidemarie.schmidt@enas.fraunhofer.de (H.S.)

Received: 28 August 2019; Accepted: 2 October 2019; Published: 11 October 2019

Abstract: In this work, we report on the impedance of p-n junction-based Si biochips with gold ring top electrodes and unstructured platinum bottom electrodes which allows for counting target biomaterial in a liquid-filled ring top electrode region. The systematic experiments on p-n junction-based Si biochips fabricated by two different sets of implantation parameters (i.e. biochips PS5 and BS5) are studied, and the comparable significant change of impedance characteristics in the biochips in dependence on the number of bacteria suspension, i.e., *Lysinibacillus sphaericus* JG-A12, in Deionized water with an optical density at 600 nm from OD_{600} = 4–16 in the electrode ring region is demonstrated. Furthermore, with the help of the newly developed two-phase electrode structure, the modeled capacitance and resistance parameters of the electrical equivalent circuit describing the p-n junction-based biochips depend linearly on the number of bacteria in the ring top electrode region, which successfully proves the potential performance of p-n junction-based Si biochips in observing the bacterial suspension. The proposed p-n junction-based biochips reveal perspective applications in medicine and biology for diagnosis, monitoring, management, and treatment of diseases.

Keywords: biochips; impedance spectroscopy; electrical equivalent circuit; biomaterial; *Lysinibacillus sphaericus* JG-A12

---

1. Introduction

Biochips [1], as one of the most advancing technologies in the biomedical field, have attracted lots of attention in the past decades due to their promising functionalities, e.g., for the detection and recognition of biomaterial in a considerable wide range [2]. In the application field of microbiology, in comparison to optical microscopy, the biochips can prevent human errors and offer faster and easier functional operation as lab measurement tools for biosensing purposes. Thus, the biochips can be helpful for the disease diagnosis with high reliability and time efficiency. Biochips possess many advantages such as mass production, simple immobilization, high density, and high throughput [3].

In this work, the miniaturized p-n junction-based Si biochips are proposed with well-defined gold ring top electrodes and unstructured platinum bottom electrodes, which offer the advantages for sensing the biomaterial such as cost-effectiveness and high portability. The impedance spectroscopy (ImS) [4] has been used to characterize the novel designed biochips. After applying the biomaterial in the Au ring top electrode region, the two-phase electrode structure has been successfully developed and investigated for establishing the functioning electrical equivalent circuit of biochips, which can be utilized for interpreting the impedance properties that recorded between the top and bottom electrodes [5]. Based on the two-phase electrode structure, the straight-forward linear relationship between the specific equivalent circuit parameters and the cell numbers has been discovered, which offers the opportunity for determining biomaterial concentration with low cost and high efficiency [6].

Furthermore, the novel p-n junction-based Si biochips possess the advantages from different perspectives. First, the analysis cost required by p-n junction-based Si biochips for determining the cell density of the biomaterials is considerably lower than that required by other methods [7]. For example, in comparison to analytical techniques such as mass spectrophotometry, gas chromatography, or liquid chromatography [8], the proposed biochips need no special treatments to the biomaterials and they can be kept alive during the detection process. In perspective applications, the number of biomaterials will be determined in a large concentration range by using p-n junction-based Si biochips in conduction with their impedance characterization. Second, the newly developed two-phase electrode structure enhanced the sensitivity of the corresponding equivalent circuit and improved the analytical accuracy of the recorded impedance properties of biochips, which enabled the possibility for sensing biomaterial with considerable low cost [9]. In this paper, the bacteria *Lysinibacillus sphaericus* JG-A12 [10] has been studied due to its potential industrial applications in metal remediation or selective recovery of metals in recycling processes. The S-layer protein in *Lysinibacillus sphaericus* JG-A12 is mainly responsible for such outstanding metal-binding capabilities. The impedance spectroscopy of p-n junction-based Si biochips may offer a new possibility for online monitoring the biomass during the cultivation process.

The paper is structured as follows: In Materials and Methods section, we describe the structure of proposed p-n junction-based Si biochips and introduce electrical equivalent circuit. In Results section, the systematical experimental study of impedance properties of biochips is demonstrated, and the equivalent circuit parameters are extracted. In Discussions section, the origin of two-phase electrode structure is studied and validated by the experimentally recorded impedance data. The paper is summarized and an outlook is given in Conclusion section.

2. Materials and Methods

The bacteria used for the investigation in the work is *Lysinibacillus sphaericus* JG-A12, i.e., a Gram-positive, rod-shaped soil bacterium isolated from the uranium mining waste pile "Haberland" near Johanngeorgenstadt in Saxony, Germany. They were cultivated in nutrient broth (8 g/L, Mast Group) overnight in Erlenmeyer flasks at 30 °C with shaking at 100 rpm. Cell density was determined by measuring the optical density at 600 nm (OD_{600}) using a UV–Vis spectrometer. Correlation between OD_{600} and cell number was achieved by cell counting under a microscope using a Neubauer counting chamber.

Detection and culture purity of the *Lysinibacillus sphaericus* JG-A12 is usually done with help of microscopy, including the morphology of colonies on agar plates, the growth behavior, even the smell of the culture. To make sure that a microorganism is *Lysinibacillus sphaericus* JG-A12, one has to use genetics means for detecting the 16S rDNA sequence. All these tests are time-consuming. The cell number could be estimated by optical density measurements (OD 600 nm) by putting culture samples into UV–VIS spectrophotometer and using established correlation between cell count by microscopy and OD values. The proposed p-n junction-based Si biochip could be used in a bypass of a culture vessel to measure cell density.

2.1. Structural Description

As illustrated in Figure 1, phosphor or boron ions have been implanted into p- or n-type silicon wafers with a thickness of 525 μm, which results in an n-p junction or a p-n junction, respectively. The 150 nm thick gold (Au) ring top electrodes have been deposited by dc-magnetron sputtering with inner and outer diameters of 6.7 mm and 8.0 mm (Figure 1b). A ring electrode has been chosen because of the homogenous field distribution between top and bottom electrodes. In the work, the biochips

PS5 and BS5 have been manufactured, measured, and modeled to analyze influence of bacteria on the biochips. Table 1 lists the overview of the implantation parameters for the manufacturing of the biochips PS5 and BS5 with ring top electrodes.

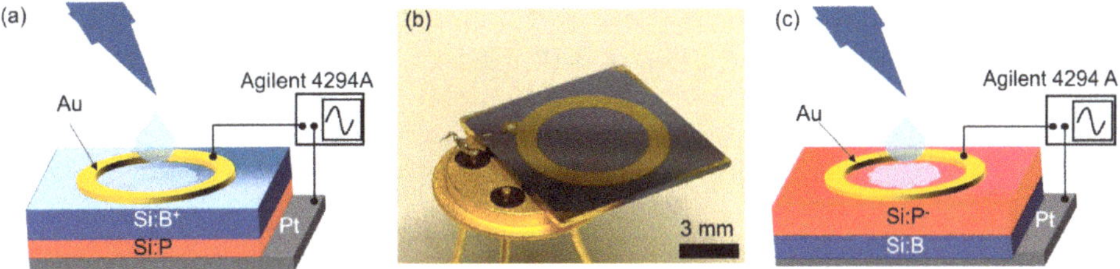

Figure 1. Schematic sketch of the p-n junction-based Si biochip with a ring top electrode with (**a**) boron ions implanted into Si:P or with (**c**) phosphorous ions implanted into Si:B. (**b**) Photograph of a socketed p-n junction-based Si biochip with Au ring electrode. Top and bottom electrodes have been wire-bonded to a diode socket and connected to an Agilent 4294A impedance bridge.

Table 1. Implantation parameters of biochips PS5 (phosphor into Si:B) and BS5 (boron into Si:P). The Au ring top electrodes and unstructured Pt bottom contacts have been prepared after ion implantation.

Biochip	Implanted Ion	Ion Energy (MeV)	Ion Fluence (cm^{-2})
PS5	Phosphor	1	3×10^{13}
BS5	Boron	0.45	3×10^{13}

The impedance characteristics of biochips PS5 and BS5 have been recorded within the frequency range from 40 Hz to 1 MHz under normal daylight at room temperature. These measurements were taken using the Agilent 4294A precision impedance analyzer. In the impedance experiments, the solvent (Deionized water) and the bacteria (*Lysinibacillus sphaericus* JG-A12) are added into the Au ring top electrode region.

In order to visualize the different concentrations of *Lysinibacillus sphaericus* JG-A12, the optical microscopic images have been taken before adding (Figure 2a) and after adding bacteria with corresponding optical density at 600 nm (OD_{600}) (Figure 2b–d). The Au ring top electrodes are deposited on a glass slide for utilizing the phase contrast mode of microscope. The OD_{600} is a common measure for microbial cell density, which can be correlated to the cell number per volume depending on the chosen bacteria. In this work, the OD_{600} of 4 up to 16 are applied in the Au ring top electrode region for further impedance characterization, which corresponds to bacteria concentration of 1.23E9 up to 6.15E9 cfu/mL under the assumption that all of the cells are alive.

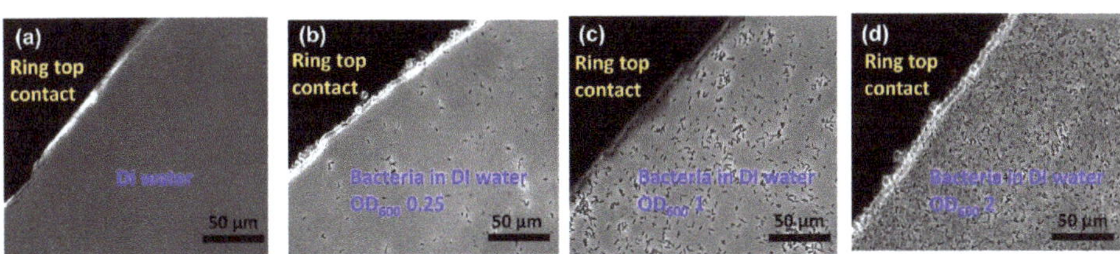

Figure 2. Top view optical microscopic image of a section of ring electrode on glass with (**a**) Deionized water, (**b**) bacteria in water at OD_{600} 0.25, (**c**) bacteria in water at OD_{600} 1, (**d**) bacteria in water at OD_{600} 2 in the ring top electrode region. Here a transparent glass substrate has been used to illuminate the sample with light from the backside. The thickness of the ring top electrodes was 150 nm and was large enough to keep the inserted liquid in the ring top electrode.

2.2. Modeling

Impedance spectroscopy (ImS) analysis is a well-established method to observe the adhesion of biomaterials because the adhesion changes the electrical behavior of the proposed biochips. Consequently, the first assumption of the electrical equivalent circuit is obtainable based on the electrical properties from the recorded Nyquist plots from biochips [11]. The complex nonlinear least square (CNLS) software is usually used to model and extract the equivalent circuit parameters from the electrical equivalent circuit. As equivalent circuit parameters, resistors or the combination of resistors and capacitors can be used to describe the ohmic or Schottky contacts in the biochips. The parallel RC pair can be used to analyze the full semicircular arc with its center on the real axis in the Nyquist plot [12]. Additionally, ImS on the proposed biochips yields imperfect semicircles with the center below the x-axis and is modeled by using constant phase elements (CPEs) [13]. The importance of CPE was highlighted by Cole and Cole [14] and was considered as an alternating current system response function [15]. CPE admittance is calculated as $Y = 1/Z = Q_0\ (j\omega)^n$, where Q_0 has the numerical value of admittance at $\omega = 1$ rad/s with the unit S. Thus, the phase angle of the CPE impedance is frequency independent and has a constant value of -(90*n) degrees. By using the CNLS software, the modeling parameters have been iteratively determined. For the CPE component, the parameters RDE (resistance), TDE (relaxation time), and PDE (phase) can be obtained. The resistance part of CPE is determined by RDE, and the capacitive part Cp in CPE can be computed as $Cp = (Q_0 * RDE)^{(1/n)}\ /RDE$, where Ω_{max} is the frequency at which $-Im\{Z\}$ is maximum on Nyquist plot and Q_0 is $Q_0 = (TDE)^{(PDE)}/RDE$. The electrical properties of the biochips can be derived from the semicircle structure of the impedance spectra in the frequency domain [16]. It has been demonstrated that the experimental impedance characteristics can be modeled by the impedance response of an electrical equivalent circuit, which consists of CPEs, resistors, capacitors, and inductors [17]. The capacitance and resistance are associated with space charge polarization regions and with particular adsorption at the electrode [18], i.e., most of the structures with electrodes normally contain a geometrical capacitance and a bulk resistance in parallel to it [19], which is the same as for the p-n junction-based Si biochips.

In the proposed p-n junction-based Si biochips, the bulk capacitance of the depletion region of the semiconductor and the capacitance of the Schottky contacts between electrodes and semiconductor contribute to the impedance spectra of the biochips. CPEs have been used to model the biochips. The electrical equivalent circuit model of the solo biochips PS5 and BS5 consist of two pairs of CPEs in parallel with resistors (Figure 3a), while the electrical equivalent circuit of the biochips after adding analytes into the Au top electrode region consists of three pairs of CPEs and resistors (Figure 3b). The equivalent circuit parameters Rs and Ls contribute to the lead impedances.

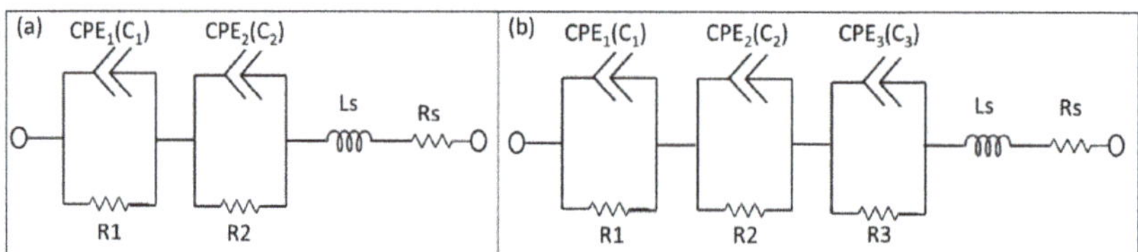

Figure 3. Electrical equivalent circuits used to model impedance spectra of the biochips PS5 and BS5 (**a**) before (two pairs of CPEs and resistors) and (**b**) after inserting analytes into the ring electrode (three pairs of CPEs and resistor). The equivalent circuit parameters Ls and Rs represent the interface properties of the circuit wiring.

3. Results

The impedance characteristics of biochips PS5 and BS5 are studied under the same experimental conditions. During the experiments, firstly, the ImS on solo biochips are measured without adding anything in the ring top electrode region, secondly, the ImS on biochips are recorded after adding 20 μL Deionized (DI) water. In the third step, for both biochips PS5 and BS5, the additional 1–5 μL DI water or bacteria suspension are applied within the ring top electrodes. Each measurement is repeated on individual biochip for 3 times, and the corresponding experimental (circular dots) and modeled results (solid lines) with error bars are shown in Figures 4 and 5 for biochip PS5 and BS5, respectively.

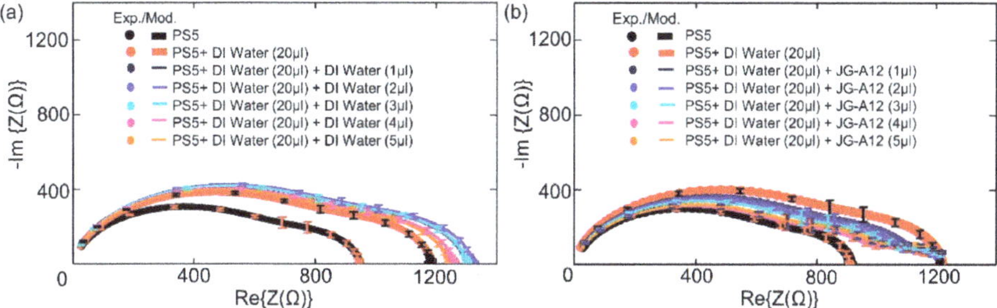

Figure 4. Experimental and modeled Nyquist plots of the Biochip PS5 with (a, b) no filling and with DI water (20 µL) and (**a**) with additional DI water with volume from 1 µL to 5 µL, and (**b**) with additional bacteria volume of JG-A12 from 1 µL to 5 µL. The error bars are inserted in all Nyquist plots according to 3 repeated Biochip experiments. The experimental results are represented in dots, and modeled results are represented in solid lines.

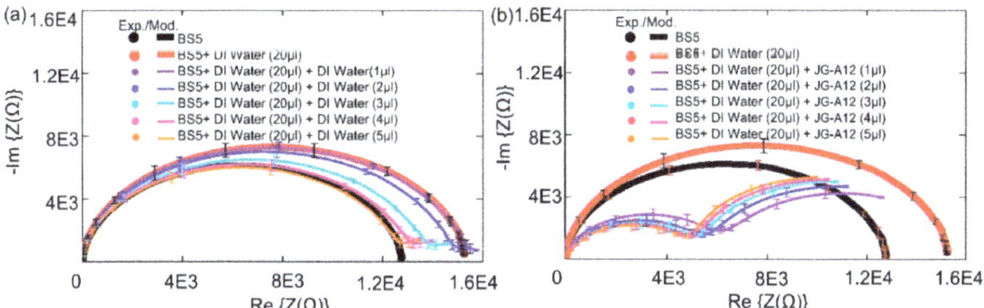

Figure 5. Experimental and modeled Nyquist plots of the Biochip BS5 with (a, b) no filling and with DI water (20 µL) and (**a**) with additional DI water volume from 1 µL to 5 µL and (**b**) with additional bacteria volume of JG-A12 from 1 µL to 5 µL. The error bars are inserted in all Nyquist plots according to 3 repeated Biochip experiments. The experimental results are represented in dots, and modeled results are represented in solid lines.

The equivalent experimental results on two PS5 biochips with the same implantation parameters without filling and with 20 µL DI water are demonstrated in Figure 4a and in Figure 4b as black and red dots, respectively, which reveals the reproducible impedance behavior of the PS5 biochips in the frequency domain. In Figure 4a, additional 1–5 µL DI water is applied after adding the 20 µL DI water, while additional 1–5 µL bacteria is applied in Figure 4b. The additionally added DI water results in an increase of corresponding capacitance and resistance parameters in the equivalent circuits in comparison to the ImS from the case after adding 20 µL DI water, whereas the additional 1–5 µL bacteria cause the decrease of both parameters. Moreover, by adding subsequential bacteria volume from 1 µL to 5 µL as shown in Figure 4b, more significant changes in the third semicircle (CPE3 and R3 as illustrated in Figure 3b) can be recorded in comparison to the impedance characteristics in Figure 4a by adding subsequential 1 µL to 5 µL DI water.

Similarly, based on the equivalent experimental results on two BS5 biochips without filling and with 20 µL DI water in the subfigures in Figure 5, the additional 1–5 µL DI water or bacteria are applied after adding the 20 µL DI water in Figure 5a,b, respectively. Note that, for biochip BS5 after adding the additional 1–5 µL DI water (Figure 5a) or 1–5 µL bacteria (Figure 5b), the decrease of corresponding capacitance and resistance parameters (the thinner curves in Figure 5) in the equivalent circuits in comparison to the ImS from the case after adding 20 μL DI water (thicker red curves in Figure 5a,b) can be recorded, which is different with the biochip PS5. Nevertheless, more significant changes from the impedance characteristics of BS5 after adding additional 1 μL–5 μL bacteria (Figure 5b) can be detected than in the case after adding additional 1–5 μL DI water (Figure 5a), which is consistent to the biochip PS5. These results successfully proved that biochips PS5 and BS5 can be used to detect adhesion of *Lysinibacillus sphaericus* JG-A12 in the ring top electrode region.

Note that the resistance of the boron-implanted biochip BS5 (Figure 5) is typically larger than the resistance of the phosphor-implanted biochip PS5 (Figure 4) due to the lower conductivity of the p-type semiconductor in which the holes are majority carriers in comparison to the n-type semiconductor whereas the electrons are the majority carriers. Conductivity is defined by σ = p.e.μh + n.e.μe, where the mobility of holes and electrons are 505 and 1450 cm^2/Vs, respectively. By adding analytes, the capacitive impedance is decreased and a third semicircle is formed. Moreover, for the BS5 biochip, the dramatic impedance variation is found after adding additional 1 μL of DI water (violet curve in Figure 5a) and 1 μL of bacterial suspension (violet curve in Figure 5b) than that of biochip PS5 (violet curves in Figure 4). The experimental impedance characteristics from the biochips PS5 can be modeled by the equivalent circuit parameters in the equivalent circuit as shown in Figure 3, which consists of two nonideal capacitances or CPEs (Cp1, Cp2), two parallel resistances (Rp1, Rp2), a contact resistance (Rs) and a contact inductance (Ls). ImS modeling of biochip PS5 is shown in Tables 2 and 3 (first row).

Table 2. Modeled equivalent circuit parameters Cp1, Rp1, Cp2, Rp2, Cp3, and Rp3 of the biochip PS5 with 20 μL DI water and different additionally inserted 1–5 μL DI water (Figure 4a).

Circuit Element	Cp1 (F)	Rp1 (Ω)	Cp2 (F)	Rp2 (Ω)	Cp3 (F)	Rp3 (Ω)	Rs (Ω)	Ls (H)
PS5	8.5×10^{-9}	6.5×10^{2}	1.4×10^{-15}	2.8×10^{6}	-	-	3.6×10^{-8}	2.4×10^{-15}
PS5+W20	8.1×10^{-9}	8.1×10^{2}	5.1×10^{-15}	9.4×10^{6}	1.4×10^{-7}	2.8×10^{2}	3.6×10^{-8}	2.4×10^{-15}
PS5+W20+W1	7.8×10^{-9}	8.7×10^{2}	5.1×10^{-15}	9.4×10^{6}	1.4×10^{-7}	3.7×10^{2}	3.6×10^{-8}	2.4×10^{-15}
PS5+W20+W2	7.6×10^{-9}	8.7×10^{2}	5.1×10^{-15}	9.4×10^{6}	1.5×10^{-7}	4.2×10^{2}	3.6×10^{-8}	2.4×10^{-15}
PS5+W20+W3	7.6×10^{-9}	8.6×10^{2}	5.1×10^{-15}	9.4×10^{6}	1.6×10^{-7}	4.4×10^{2}	3.6×10^{-8}	2.4×10^{-15}
PS5+W20+W4	7.6×10^{-9}	8.3×10^{2}	5.1×10^{-15}	9.4×10^{6}	1.6×10^{-7}	4.4×10^{2}	3.6×10^{-8}	2.4×10^{-15}
PS5+W20+W5	7.6×10^{-9}	8.1×10^{2}	5.1×10^{-15}	9.4×10^{6}	1.7×10^{-7}	4.2×10^{2}	3.6×10^{-8}	2.4×10^{-15}

W20 = 20 μL DI water, W1 = 1 μL DI Water, W2 = 2 μL DI Water, W3 = 3 μL DI Water, W4 = 4 μL DI Water, W5 = 5 μL DI Water.

Table 3. Modeled equivalent circuit parameters Cp1, Rp1, Cp2, Rp2, Cp3, and Rp3 of the biochip PS5 and with 20 μL DI water and additionally inserted 1–5 μL bacteria (Figure 4b).

Circuit Element	Cp1 (F)	Rp1 (Ω)	Cp2 (F)	Rp2 (Ω)	Cp3 (F)	Rp3 (Ω)	Rs (Ω)	Ls (H)
PS5	8.5×10^{-9}	6.5×10^{2}	5.1×10^{-15}	9.4×10^{6}	-	-	3.6×10^{-8}	2.4×10^{-15}
PS5+W20	8.1×10^{-9}	8.1×10^{2}	5.1×10^{-15}	9.4×10^{6}	1.4×10^{-7}	2.8×10^{3}	3.6×10^{-8}	2.4×10^{-15}
PS5+W20+B1	2.1×10^{-9}	7.8×10^{2}	5.1×10^{-15}	9.4×10^{6}	7.8×10^{-8}	2.4×10^{3}	3.6×10^{-8}	2.4×10^{-15}
PS5+W20+B2	2.2×10^{-9}	7.5×10^{2}	5.1×10^{-15}	9.4×10^{6}	8.2×10^{-8}	2.4×10^{3}	3.6×10^{-8}	2.4×10^{-15}
PS5+W20+B3	2.4×10^{-9}	7.3×10^{2}	5.1×10^{-15}	9.4×10^{6}	8.8×10^{-8}	2.2×10^{3}	3.6×10^{-8}	2.4×10^{-15}
PS5+W20+B4	2.5×10^{-9}	6.9×10^{2}	5.1×10^{-15}	9.4×10^{6}	9.0×10^{-8}	2.1×10^{3}	3.6×10^{-8}	2.4×10^{-15}
PS5+W20+B5	2.7×10^{-9}	6.7×10^{2}	5.1×10^{-15}	9.4×10^{6}	10×10^{-8}	2.1×10^{3}	3.6×10^{-8}	2.4×10^{-15}

W20 = 20 μL DI water, B1 = 1 μL bacteria, B2 = 2 μL bacteria, B3 = 3 μL bacteria, B4 = 4 μL bacteria, B5 = 5 μL bacteria.

It should be noted that the equivalent circuit of the impedance spectra of the solo biochip PS5 and biochip PS5 with analytes are different due to the additional appeared semicircle. Based on the experimental impedance characteristics from the biochips PS5 after adding analytes, the composition and cell numbers of analytes added to the Au top ring electrode region for biochip PS5 can be determined by modeling the equivalent circuit parameters in the equivalent circuit as shown in Figure 3, which consists of three nonideal capacitors (Cp1, Cp2, Cp3) and three resistors (Rp1, Rp2, Rp3), a contact resistance (Rs), and a contact inductance (Ls). The corresponding ImS modeling results of the biochip PS5 are shown in Table 2 (with DI water) and in Table 3 (with DI water and bacteria).

Similarly, the ImS characteristics of the biochip BS5 can be modeled with two pairs of resistors and nonideal capacitors (Rp1, Rp2, Cp1, Cp2), contact resistance (Rs), and contact inductance (Ls). Furthermore, for the biochip BS5 the electrical equivalent circuit model for biochip with analytes consists of three pairs of resistance-capacitance in addition to the contact resistance and contact inductance. The corresponding ImS modeling results of the biochip BS5 are shown in Table 4 (with DI water) and in Table 5 (with bacteria).

Table 4. Modeled equivalent circuit parameters Cp1, Rp1, Cp2, Rp2, Cp3, and Rp3 of the biochip BS5 with 20 µL DI water and additionally inserted 1–5 µL DI water (Figure 5a).

Circuit Element	Cp1 (F)	Rp1 (Ω)	Cp2 (F)	Rp2 (Ω)	Cp3 (F)	Rp3 (Ω)	Rs (Ω)	Ls (H)
BS-5	4.6×10^{-9}	1.2×10^{4}	7.5×10^{-15}	9.4×10^{6}	-	-	2.1×10^{-9}	2.4×10^{-15}
BS5+W20	4.6×10^{-9}	1.5×10^{4}	7.5×10^{-15}	9.4×10^{6}	2.6×10^{-6}	6.2×10^{5}	2.1×10^{-9}	2.4×10^{-15}
BS5+W20+W1	4.5×10^{-9}	1.5×10^{4}	7.5×10^{-15}	9.4×10^{6}	2.9×10^{-5}	2.1×10^{3}	2.1×10^{-9}	2.4×10^{-15}
BS5+W20+W2	4.7×10^{-9}	1.4×10^{4}	7.5×10^{-15}	9.4×10^{6}	3.5×10^{-5}	2.3×10^{3}	2.1×10^{-9}	2.4×10^{-15}
BS5+W20+W3	4.8×10^{-9}	1.3×10^{4}	7.5×10^{-15}	9.4×10^{6}	3.9×10^{-5}	2.5×10^{3}	2.1×10^{-9}	2.4×10^{-15}
BS5+W20+W4	4.8×10^{-9}	1.2×10^{4}	7.5×10^{-15}	9.4×10^{6}	4.2×10^{-5}	4.3×10^{3}	2.1×10^{-9}	2.4×10^{-15}
BS5+W20+W5	4.9×10^{-9}	1.2×10^{4}	7.5×10^{-15}	9.4×10^{6}	4.5×10^{-5}	5.0×10^{3}	2.1×10^{-9}	2.4×10^{-15}

W20 = 20 µL DI water, W1 = 1 µL DI water, W2 = 2 µL DI water, W3 = 3 µL DI water, W4 = 4 µL DI water, W5 = 5 µL DI water.

Table 5. Modeled equivalent circuit parameters Cp1, Rp1, Cp2, Rp2, Cp3, and Rp3 of the biochip BS5 and with 20 µL DI water and additionally inserted 1–5 µL bacteria (Figure 5b).

Circuit Element	Cp1 (F)	Rp1 (Ω)	Cp2 (F)	Rp2 (Ω)	Cp3 (F)	Rp3 (Ω)	Rs (Ω)	Ls (H)
BS5	4.6×10^{-9}	1.2×10^{4}	7.5×10^{-15}	9.4×10^{6}	-	-	2.1×10^{-9}	2.4×10^{-15}
BS5+W20	4.6×10^{-9}	1.5×10^{4}	7.5×10^{-15}	9.4×10^{6}	2.6×10^{-6}	6.2×10^{5}	2.1×10^{-9}	2.4×10^{-15}
BS5+W20 +B1	9.7×10^{-9}	6.1×10^{3}	7.5×10^{-15}	9.4×10^{6}	5.6×10^{-7}	1.0×10^{4}	2.1×10^{-9}	2.4×10^{-15}
BS5+W20 +B2	1.1×10^{-8}	5.4×10^{3}	7.5×10^{-15}	9.4×10^{6}	7.0×10^{-7}	1.1×10^{4}	2.1×10^{-9}	2.4×10^{-15}
BS5+W20 +B3	1.4×10^{-8}	5.2×10^{3}	7.5×10^{-15}	9.4×10^{6}	7.2×10^{-7}	1.1×10^{4}	2.1×10^{-9}	2.4×10^{-15}
BS5+W20 +B4	1.7×10^{-8}	5.0×10^{3}	7.5×10^{-15}	9.4×10^{6}	7.6×10^{-7}	1.2×10^{4}	2.1×10^{-9}	2.4×10^{-15}
BS5+W20 +B5	1.8×10^{-8}	4.8×10^{3}	7.5×10^{-15}	9.4×10^{6}	7.8×10^{-7}	1.2×10^{4}	2.1×10^{-9}	2.4×10^{-15}

W20 = 20 µL DI water, B1 = 1 µL bacteria, B2 = 2 µL bacteria, B3 = 3 µL bacteria, B4 = 4 µL bacteria, B5 = 5 µL bacteria

With the help of optical microscopy, the calibration between the ImS data and cell concentration observed with the optical density at 600 nm (OD_{600}) has been considered. Calibration of the biochip is achieved in the volume range from 0 µL to 5 µL bacterial suspension in 20 µL DI water. OD_{600} of 4 corresponds to 2.46E7 cells on the chip if *Lysinibacillus sphaericus* JG-A12 with 1 µL of concentration is applied to 20 µL DI water. For calibration, the dependency of the modeled equivalent circuit elements Rp1, Rp2, Cp1, Cp2 and Rp3 and Cp3 (from impedance modeling) on the nominal number of bacterial cells (from optical microscopy) was evaluated on the basis of the biochip PS5 (Figure 6).

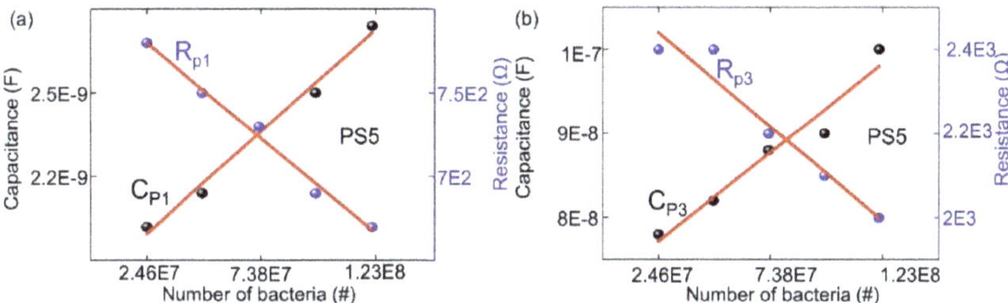

Figure 6. Modeled equivalent circuit parameters (dots) and linear fitting curve (red lines) for (**a**) Rp1, and Cp1 and for (**b**) Rp3 and CP3 of the biochip PS5 in dependence on the number of *Lysinibacillus sphaericus* JG-A12.

As demonstrated in Figure 6, 4 equivalent circuit parameters Rp1, Cp1, Rp3, and Cp3 from biochip PS5 have been proved to possess the linear dependence with the number of bacteria. Moreover, 3 equivalent circuit parameters Cp1, Rp3, and Cp3 have the linear relationship with the nominal number of bacterial cells for the biochip BS5 in the range from 2.46E7 to 1.23E8 (Figure 7).

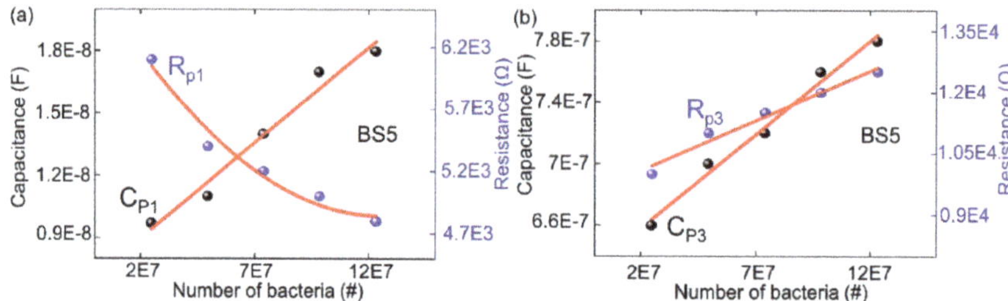

Figure 7. Interpolation (red line) of modeled equivalent circuit parameters (**a**) Rp1 and Cp1 and of (**b**) Rp3 and CP3 of the Si Biochip BS5 that depend on number of *Lysinibacillus sphaericus* JG-A12.

Among the modeled electrical elements in the equivalent circuit in Figure 3b, the Rp1 and Cp1 pair represents the Schottky contact at the electrodes/semiconductor interface. If the size of contact area is denoted as A, by adding bacteria suspension to the top electrode region of biochips, the area of the top contact is increased. According to the equation Rp1 = ρ(d/A), where d denotes the thickness of the Schottky barrier, the resistance is reversely related to the area A. Thus, there is reduction in resistance by adding the bacteria suspension. If we consider Cp1 = ε(A/d) with ε as the permittivity of semiconductor, the relationship between Cp1 and A results in the increasing Cp1 with increasing bacteria suspension. The Rp2 and Cp2 pair corresponds to the impedance of semiconductors Si:B in PS5 and Si:P in BS5, and the values are kept the same in Table 2/Table 3 for PS5 and in Table 4/Table 5 for BS5, respectively. The Rp3 and Cp3 pair represents the impedance of bacterial suspension which is added into the Au top electrode region. According to the experimental results from biochip PS5, Rp3 is decreasing linearly with bacteria concentration, while BS5 Rp3 is increasing linearly. Cp3 is increased linearly in both cases.

In detail, the linear impedance change that depends on the bacterial concentration for the Si Biochip PS5 is illustrated in Figure 6 with the 4 model parameters Rp1, Cp1, Rp3, and CP3. Furthermore, the linear impedance change that depends on the bacterial concentration for Si Biochip BS5 is illustrated in Figure 7 with the 3 model parameters Cp1, Rp3, and CP3. Therefore, a multiparameter determination of the bacterial concentration can be performed for both Si Biochips.

4. Discussion

Due to the appearance of non-overlapping semicircular curves in the Nyquist plots which have been shown in Figures 4 and 5, it can be directly estimated that the associated resistor (R) and capacitor (C) can be used for describing the physical structures of the biochips [20]. Based on this initial estimation and the experimental impedance data (Figure 8), the transferring of the equivalent circuits, based on the physical mechanism of the biochips (Figure 9a,c) to the ones based on the assumed associated resistor-capacitor RC pairs (Figure 9b,d), can be validated.

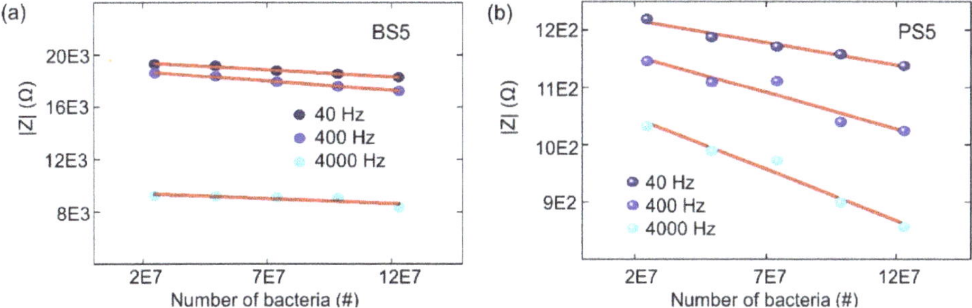

Figure 8. Impedance magnitude measured at test frequencies of 40 Hz, 400 Hz, and 4000 Hz on biochips (**a**) BS5 (Table S1) and (**b**) PS5 (Table S2) that depend on the number of bacteria. The experiments are carried out under normal illumination.

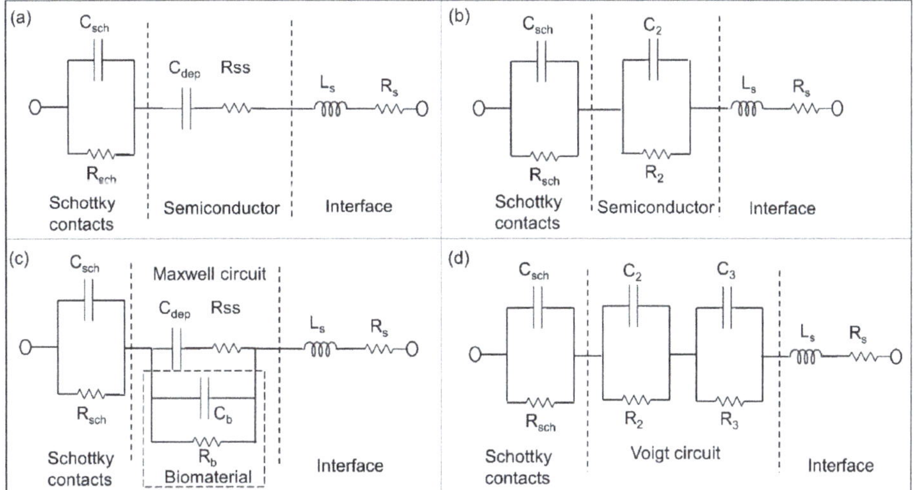

Figure 9. Equivalent circuit models of biochips without analyte bacteria (**a**) based on the physical structure of biochip and (**b**) based on the associated RC pairs. The parallel capacitor C_2 and resistor R_2 are transferred from C_{dep} and Rss. Equivalent circuit models of biochip with analyte (**c**) in Maxwell fashion and in (**d**) Voigt fashion. The Voigt fashion can be transferred into Maxwell fashion by using Equations (1)–(3).

For the proposed p-n junction-based Si biochip, the Schottky contacts are formed at the interfaces both between Au top electrode/semiconductor and between Pt bottom electrode/semiconductor. These two metal/semiconductor Schottky contacts can be represented by one pair of CPE and resistor in the electrical equivalent circuit [21]. As demonstrated in Figure 9a, the p-n junction in the biochip PS5 or BS5 can be described by depletion region capacitor C_{dep} and semiconductor resistor Rss, which can be further converted to the parallel C_2 // R_2 pair fashion (Figure 9b), where $C_2 = C_{dep} \cdot (Q^2/(1+Q^2))$, $R_2 = R_{ss} \cdot (1+Q^2)$ with the definition of $Q = 1/(\omega \cdot C_{dep} \cdot Rss)$ [22]. Thus, the impedance spectra from solo biochips PS5 and BS5 as depicted in thick black curves in Figures 4 and 5 can be modeled by using equivalent circuits in Figure 9b.

After applying the analytes (DI water or bacteria), a two-phase electrode contact [23] can be developed by using the proposed biochips, where one phase is resulted from the electrodes of biochips and the other phase is related to the added analytes [24]. The Maxwell circuit in Figure 9c can be transferred into Voigt circuit as illustrated in Figure 9d by utilizing the following equations [25]:

$$C_{2,3} = 2C_b \left(1 \mp \frac{\frac{Rss}{Rb} - \frac{Cb}{Cdep+1}}{k^{1/2}} \right)^{-1}, \tag{1}$$

$$R_{2,3} = \frac{Rb}{2} \left(1 \pm \frac{\frac{Cb}{Cdep} - \frac{Rss}{Rb+1}}{k^{1/2}} \right), \tag{2}$$

$$k = \left(\frac{Cb}{Cdep} + \frac{Rss}{Rb} + 1 \right)^2 - \frac{4Cb.Rss}{Cdep.Rb}. \tag{3}$$

The validity of equivalent circuit for biochips after applying bacteria can be described by experimental impedance data as shown in Figure 8. Here, the impedance magnitude of the biochips PS5 and BS5 after adding additional 1–5 µL bacteria has been depended on cell concentration. The overall impedance magnitude at different frequencies is decreasing with increasing bacteria concentration, which indicates the validity of parallel connection of RC pairs for the equivalent circuit of biochips with added bacteria. It means that the RC pair which is corresponding to the added bacteria should be in parallel with the p-n junction-based RC pair as shown in Figure 9d. Thus, it can be concluded that the additional semicircles after adding the bacteria as demonstrated in Figures 4b and 5b are caused by the liquid phase in the two-phase electrode structure.

Based on the development of two-phase electrode contact, the final equivalent circuit for the biochips with analytes can be achieved as shown in Figure 3b, where three RC pairs applied. The two-phase electrode structure offers the opportunity for detecting the adhesion of bacteria by analyzing the impedance changes recorded in the Nyquist plot with higher accuracy and efficiency.

5. Conclusions

In this work, we propose the p-n junction-based Si biochips with gold ring top electrodes and unstructured platinum bottom electrodes, which offer a novel possibility for online monitoring the biomass during the cultivation with considerable low cost. The promising reproducible impedance characteristics of two types of p-n junction-based Si biochips, i.e., PS5 and BS5, with different implantation ion types and ion energy but the same ion fluence, are discussed and demonstrate the significant change between the impedance spectra before and after adding the bacteria suspension *Lysinibacillus sphaericus* JG-A12 in the top electrode region. With the help of the developed two-phase electrode structure, the equivalent circuit is designed for the p-n junction-based Si biochips, four modeling parameters for PS5 (Rp1, Cp1, Rp3, and Cp3) and three modeling parameters for BS5 (Cp1, Rp3, and Cp3) reveal the linear dependence relationship with the bacterial concentration. Such linear dependent parameters are useful for the quantitative measure of the bacteria concentration. The limitation of the bacteria detection utilizing p-n junction-based Si biochips strongly depends on inner and outer diameters of the ring top electrode. Thus, for detecting a smaller number of bacteria, the inner and outer diameters should be reduced. In the future, the p-n junction-based Si biochips by adding different live cells or tissues with different cell dimensions will be studied and their further potential in medicine and biology applications will be explored.

Supplementary Materials: The following are available online at http://www.mdpi.com/2079-6374/9/4/120/s1. Table S1: The impedance magnitude for boron-doped biochip BS5 with different amounts of analyte (1–5 µL bacteria) at frequencies 40 Hz, 400 Hz, and 4 kHz. This table corresponds to the experimental results in Figure 8a. Table S2. The impedance magnitude for the biochip PS5 with different amounts of analyte (1–5 µL bacteria) at frequencies 40 Hz, 400 Hz, and 4 kHz. This table corresponds to the experimental results in Figure 8b.

Author Contributions: Conceptualization: H.S. and J.R.; methodology: H.S., N.D., and D.B.; software: M.K. and N.D.; validation: M.K.; formal analysis: M.K., N.D., and H.S.; investigation: M.K. and N.D.; resources: U.H., M.V., I.S., and D.B.; data curation: M.K.; writing—original draft preparation, M.K.; writing—review and editing, N.D.

and H.S.; visualization: J.R.; supervision: H.S. and N.D.; project administration: H.S., S.E.S., and O.G.S.; funding acquisition: H.S., S.E.S., and O.G.S.

Funding: Funding by Sächsische Aufbaubank (SAB) is gratefully acknowledged (PolCarr-Sens project, grant number: 100260515).

Conflicts of Interest: The authors declare no conflict of interest.

References

1. D'Souza, S.F. Microbial biosensors. *Biosens. Bioelectron.* **2001**, *16*, 337–353. [CrossRef]
2. Aloisi, A.; Della Torre, A.; De Benedetto, A.; Rinaldi, R. Bio-Recognition in Spectroscopy-Based Biosensors for *Heavy Metals-Water and Waterborne Contamination Analysis. *Biosensors* **2019**, *9*, 96. [CrossRef] [PubMed]
3. Neves, M.M.P.S.; Martín-Yerga, D. Advanced Nanoscale Approaches to Single-(Bio)entity Sensing and Imaging. *Biosensors* **2018**, *8*, 100. [CrossRef] [PubMed]
4. Zhang, F.Z.; Keasling, J. Biosensors and their applications in microbial metabolic engineering. *Trends Microbiol.* **2011**, *19*, 323–329. [CrossRef] [PubMed]
5. Arefev, K.M.; Guseva, M.A.; Khomchenkov, B.M. Determination of the coefficient of diffusion of cadmium and magnesium vapors in gases by the Stefan method. *High Temp.* **1987**, *25*, 174–179.
6. Selenska-Pobell, S.; Panak, P.; Miteva, V.; Boudakov, I.; Bernhard, G.; Nitsche, H. Selective accumulation of heavy metals by three indigenous Bacillus strains, B. cereus, B. megaterium and B. sphaericus, from drain waters of a uranium waste pile. *FEMS Microbiol. Ecol.* **1999**, *29*, 59–67. [CrossRef]
7. Su, L.; Jia, W.; Hou, C.; Lei, Y. Microbial biosensors: A review. *Biosens. Bioelectron.* **2011**, *26*, 1788–1799. [CrossRef] [PubMed]
8. Shul'ga, A.A.; Soldatkin, A.P.; El'skaya, A.V., Dzyadevich, S.V.; Patskovsky, S.V.; Strikha, V.I. Thin-film conductometric biosensors for glucose and urea determination. *Biosens. Bioelectron.* **1994**, *9*, 217–223. [CrossRef]
9. Durrieu, C.; Lagarde, F.; Jaffrezic-Renault, N. Nanotechnology assets in biosensors design for environmental monitoring. In *Nanomaterials: A Danger or a Promise*; Brayner, R., Fiévet, F., Coradin, T., Eds.; Springer: London, UK, 2013.
10. Raff, J.; Soltmann, U.; Matys, S.; Selenska-Pobell, S.; Böttcher, H.; Pompe, W. Biosorption of Uranium and Copper by Biocers. *Chem. Mater.* **2003**, *15*, 240–244. [CrossRef]
11. Scognamiglio, V. Nanotechnology in glucose monitoring: Advances and challenges in the last 10 years. *Biosens. Bioelectron.* **2013**, *47*, 12–25. [CrossRef] [PubMed]
12. Molero, F.J.; Crespo, F.; Ferrer, S. A Note on Reparametrizations of the Euler Equations. *Theory Dyn. Syst.* **2017**, *16*, 453. [CrossRef]
13. Fricke, H. The theory of Electrolyte Polarization. *Philos. Mag.* **1953**, *14*, 310–318. [CrossRef]
14. Kerrison, L. The capacitor. *Stud. Q. J.* **1933**, *4*, 100. [CrossRef]
15. Cole, K.S.; Cole, R.H. Dispersion and Absorption in Dielectrics.I. Alternating-Current Characteristics. *J. Chem. Phys.* **1941**, *9*, 341–351. [CrossRef]
16. McKubre, M.C.H.; Macdonald, D.D. Electronic Instrumentation for Electrochemical Studies. In *A Comprehensive Treatise of Electrochemistry*; Bockris, J.O., Conway, B.E., Yeager, E., Eds.; Plenum Press: Boston, MA, USA; New York, NY, USA, 1984.
17. Zandman, F.; Stein, S. A New Precision Film Resistor Exhibiting Bulk Properties. *IEEE T. Compon. Pack T.* **1964**, *11*, 107–119. [CrossRef]
18. Usui, Y. Equivalent Circuit of Distributed Constant Circuit and Transmission Equation. *JIEP.* **2018**, *21*, 311–316. [CrossRef]
19. Van Putten, A.F.P. Interfacing to Sensors. In *Electronic Measurement Systems*; CRC Press: Boca Raton, FL, USA, 1996. [CrossRef]
20. Karuthedath, C.; Schwesinger, N. Unique and Unclonable Capacitive Sensors. *IEEE Sens. J.* **2018**, *18*, 6097–6105. [CrossRef]
21. Sihvola, A. *Electromagnetic Mixing Formulas and Applications*; The Institute of Electrical Engineers: London, UK, 1999.
22. Bauerle, J.E. Study of Solid Electrolyte Polarization by a Complex Admittance Method. *J. Phys. Chem. Solids* **1969**, *30*, 2657–2670. [CrossRef]

23. Fricke, H. The Maxwell-Wagner Dispersion in a Suspension of Ellipsoids. *J. Phys. Chem.* **1932**, *57*, 934–937. [CrossRef]
24. Barsoukov, E.; Ross Macdonald, J. *Impedance Spectroscopy: Theory, Experiment, and Applications*; John Wiley & Sons: Hoboken, NJ, USA, 2018.
25. Novoseleskii, M.; Gudina, N.N.; Fetistov, Y.I. Identical Equivalent Impedance Circuits. *Sov. Electrochem.* **1972**, *8*, 546–548.

 biosensors

Article

Carbon Nanotube-Based Electrochemical Biosensor for Label-Free Protein Detection

Jesslyn Janssen [1,*], Mike Lambeta [2,3], Paul White [3] and Ahmad Byagowi [3]

1 Department of Biological Sciences, University of Manitoba, Winnipeg, MB R3T 2N2, Canada
2 Department of Physics and Astronomy, University of Manitoba, Winnipeg, MB R3T 2N2, Canada; mike@lambeta.net
3 Department of Electrical and Computer Engineering, University of Manitoba, Winnipeg, MB R3T 2N2, Canada; pauljwhite@gmail.com (P.W.); ahmadexp@gmail.com (A.B.)
* Correspondence: janssen7@berkeley.edu

Received: 3 October 2019; Accepted: 9 December 2019; Published: 17 December 2019

Abstract: There is a growing need for biosensors that are capable of efficiently and rapidly quantifying protein biomarkers, both in the biological research and clinical setting. While accurate methods for protein quantification exist, the current assays involve sophisticated techniques, take long to administer and often require highly trained personnel for execution and analysis. Herein, we explore the development of a label-free biosensor for the detection and quantification of a standard protein. The developed biosensors comprise carbon nanotubes (CNTs), a specific antibody and cellulose filtration paper. The change in electrical resistance of the CNT-based biosensor system was used to sense a standard protein, bovine serum albumin (BSA) as a proof-of-concept. The developed biosensors were found to have a limit of detection of 2.89 ng/mL, which is comparable to the performance of the typical ELISA method for BSA quantification. Additionally, the newly developed method takes no longer than 10 min to perform, greatly reducing the time of analysis compared to the traditional ELISA technique. Overall, we present a versatile, affordable, simplified and rapid biosensor device capable of providing great benefit to both biological research and clinical diagnostics.

Keywords: electrochemical biosensors; SWCNT; point-of-care diagnostics; label-free biosensors; ELISA; carbon nanotubes; bovine serum albumin

---

1. Introduction

Recently, the need for affordable, rapid and simplified diagnostic devices has been growing, especially in third world countries. Many diseases such as cancer and infectious disease require an early-stage diagnosis, when treatment options can be most effective. It is also suspected that drug discovery may be accelerated for incurable diseases such as Alzheimer's disease where the inability to diagnose the disease at its earliest stages may be leading to ineffective drug trials which are usually administered to patients that possess irreversible, late-stage pathology. In the case of breast cancer, 5-year survival rates are much lower for developing countries such as Gambia (12.0%) and Algeria (38.8%) in comparison to the United States of America (83.9%) and Sweden (82.0%) [1]. It is believed that lower survival rates in developing countries are due to diagnosis in the advanced stages of disease and access barriers to medical care [2–4]. In the case of infectious disease, just under 1 million people die from malaria, 4.3 million people from acute respiratory infections and 2.9 million from enteric infections every year [5]. More than 90% of the deaths due to infectious disease occur in developing countries [5]. Affordable, rapid and simplified diagnostics are critical for combatting these diseases, yet most diagnostic methods are inaccessible to those who need them most.

Currently, immunological assays such as enzyme-linked immunosorbent assay (ELISA) for the detection of biomarkers in bodily fluids are sensitive and provide accurate results. However, the current assays involve sophisticated techniques and require highly trained personnel for execution and analysis. It may take days to weeks for patients to obtain results from the current immunoassay technologies. ELISA also requires large amounts of samples and due to its label-based approach, expensive and specialized reagents are needed, which prevents the use of these assays in resource-poor environments [6,7]. Therefore, the need for a rapid, inexpensive and simplified diagnostic device is still unmet.

Biosensing technology is a promising alternative, owing to its potential for rapid, simple, sensitive, low-cost and portable detection [8]. Biosensors are devices that comprise a biological component for recognition, and a physiochemical detector component for transduction [9]. The transduction component can be optical, electrochemical, piezoelectric, magnetic or calorimetric [9]. The biological recognition component can be developed using enzymes, antibodies, cells, tissues, peptides, nucleic acids and aptamers [10–14]. In regard to biosensing technology for protein detection, there is an interest in label-free biosensors as they require only a single recognition element, leading to a simplified design and a reduction in reagent costs and assay time [15]. Optical interferometry [16], fiber optic surface plasmon resonance [17–20], piezoelectric [21–24] and electrochemical [25–28] label-free methods have been developed to overcome the limitations of label-based biosensing technologies. In particular, electrochemical label-free methods are most promising with regard to high sensitivity, lower detection limit, lower response time, cost-effectiveness, miniaturization, simplification and portability [29,30]. Electrochemical measurement methods are most suitable for mass fabrication and have played a crucial role in the transition towards simplified point-of-care diagnostics [11]. This shift has been evident through the market domination of self-testing glucose strips, based on screen-printed enzyme electrodes, coupled to pocket-sized amperometric meters for diabetes over the past two decades [31]. Although various research efforts have been made, the development of a simplified, affordable, easy-to-use and highly performant protein detection system for point-of-care analysis remains a challenge.

Electrochemical biosensors that utilize nanomaterials such as carbon nanotubes (CNTs) for improved sensitivity and response time are potential candidates for point-of-care protein detection. CNTs are hollow, cylindrical molecules consisting of a hexagonal arrangement of hybridized carbon atoms, one or more walls and a nanometer scale diameter [32]. Their well-ordered arrangement of carbon atoms is linked via sp^2 bonds, making them the stiffest and strongest fibers known [32]. Depending on their number of walls, CNTs can be divided into single-walled carbon nanotubes (SWCNTs) and multi-walled carbon nanotubes. In particular, SWCNTs offer great promise for biosensing applications due to a unique combination of electrical, magnetic, optical, mechanical and chemical properties [32]. SWCNTs exhibit the simplest morphology and may be formed by rolling up a single graphene sheet. The large surface area of SWCNTs enables interactions with a large number of biological agents at the carbon nanotube surface, allowing for sensitive detection of biomolecular interactions [33]. Additionally, the electrical conductance of SWCNTs are sensitive to the environment and vary significantly with surface adsorption of various chemicals and biomolecules, making SWCNTs promising candidates for label-free biosensing [34–36].

Due to their unique characteristics, SWCNTs have been used in several label-free detection modalities. One such label-free modality includes electrical percolation-based biosensors. In electrical percolation, a long-range electrical connectivity is formed in randomly oriented and distribute systems of conductive elements, such as SWCNTs [37]. In these systems, the passage of current through the conductive network depends on the continuity of the network [36]. Electrical percolation-based biosensors often comprise a 3-dimensional carbon nanotube and antibody network [37]. In such biosensors, molecular interactions, such as the binding of antigens to antibodies, disrupt the network continuity causing an increase in the network's electrical resistance [37].

This paper describes the development and performance of a label-free disposable biosensor based on cellulose paper impregnated with SWCNTs and antibodies for protein detection. The biosensors take advantage of the electrical percolation principle in order to facilitate label-free and simplified detection of protein biomarkers. Biosensors were developed for the measurement of BSA and comprised an antibody specific to BSA. These biosensors were developed for the detection and quantification of BSA because it is often used as a protein concentration standard in lab experiments. Additionally, targeting such a protein could be useful for determining whether our developed biosensors were capable of measuring a standard protein in solution as a proof-of-concept and whether our method could compete with other standard techniques such as ELISA. The data obtained in this work indicate that such biosensors may be utilized as a foundation for rapid, inexpensive and simplified diagnostics.

2. Materials and Methods

2.1. *Material and Reagents*

The poly(sodium 4-styrene sulfonate) (PSS) and antibodies specific to bovine serum albumin were obtained from Sigma Aldrich (St. Louis, MO, USA). SWCNTs were purchased from Cheap Tubes Inc. (Brattleboro, VT, USA) and used without purification. Filter paper discs were cut from Whatman cellulose filtration paper obtained from Sigma Aldrich (St. Louis, MO, USA). A bovine serum albumin stock solution bought from Thermo Fisher Scientific (Rockford, IL, USA) was used for the preparation of various testing concentrations of BSA. Ovalbumin stock solution was purchased from Sigma Aldrich (St. Louis, MO, USA). The BSA ELISA kit was purchased from Cygnus Technologies (Southport, NC, USA).

2.2. *Biosensor Fabrication*

SWCNTs and PSS were dispersed in distilled water to achieve a final concentration of 50 mg/mL. The SWCNT and PSS solution underwent bath sonication for 30 min prior to centrifugation for 20 min for the removal of aggregates. PSS was used to facilitate electrostatic adsorption for protein immobilization. 8 µL of BSA antibody was added to 5 mL of the PSS and SWCNT solution. The SWCNT-antibody solution was vortexed for 1 min and then drop-casted onto 3 mm × 3 mm cellulose paper disks. The disks were then placed onto a printed circuit board (PCB) electrode and freeze-dried under vacuum to prevent antibody denaturation. Biosensors were also developed without the use of BSA antibody, in order to control for non-specific BSA binding.

2.3. *Sample Preparation*

Various concentrations of BSA were prepared from a 10% BSA stock solution. 1 ng/mL, 5 ng/mL, 10 ng/mL, 20 ng/mL and 40 ng/mL BSA solutions were prepared. Ovalbumin standard solutions used in our control experiment were prepared from a 10% stock solution in a similar manner.

2.4. *Measurements*

5 µL of each BSA standard solution were placed onto a biosensor disk and the biosensors were left to dry. To determine the optimal temperature at which the biosensors should dry, or the immunoreaction should occur, the drying and subsequent measurements were carried out under various temperatures (4 °C, 23 °C, 37 °C and 50 °C). All electrical measurements were performed using a programmed Arduino Uno. The Arduino measurement device was programmed to measure electrical resistance and record the value at the transient plateau of resistance readings, before subtracting the recorded value from the initial measurement that was taken prior to the reaction, resulting in a final value for the change in resistance (see supplementary online material for further details). Changes in resistance were correlated with the concentration of BSA. To control for non-specific binding, various concentrations of BSA were applied to biosensors developed without the use of the BSA antibody. Additionally, concentrations of a non-target analyte solution were applied to the developed biosensors. The standard solutions of BSA were analyzed using traditional ELISA for verification and comparison. All experiments were carried out five times before the average values were calculated. The detection limit was calculated using LOD = 3.3 SD/S, where LOD is the limit of detection, SD is the standard deviation and S is the slope of the resulting standard curve of the trials with BSA standard solutions.

3. Results and Discussion

3.1. Biosensor Design

Accurate and rapid quantification of protein biomarkers plays an important role in the diagnosis of numerous diseases. With the potential that carbon nanotube biosensors offer for rapid, simple, sensitive and portable diagnostics, we developed SWCNT-coated paper-based biosensors for the detection and quantification of a standard protein, bovine serum albumin. Similar to a finger-prick blood test, the developed SWCNT biosensors were designed for the quantification of a specific protein in a single drop (5 μL) of sample solution.

Only SWCNTs were used for the fabrication of the biosensors, due to their larger surface area. Cellulose paper discs were drop-casted with the SWCNT/PSS/antibody solution, 6 times in a row to achieve an electrical resistance in a sufficient 200–250 Ω range. The discs were placed onto a printed circuit board (PCB) electrode which was subsequently freeze dried under vacuum. One of the fabricated biosensors is depicted in Figure 1. All biosensors used during experimentation had SWCNT-coated discs that were firmly attached to the PCB substrate. Attaching the SWCNT-coated discs to the PCB electrodes in this way allowed us to avoid clamping electrodes that bend the flexible paper-based biosensors, causing resistance changes due to strain that interfere with analyte measurement.

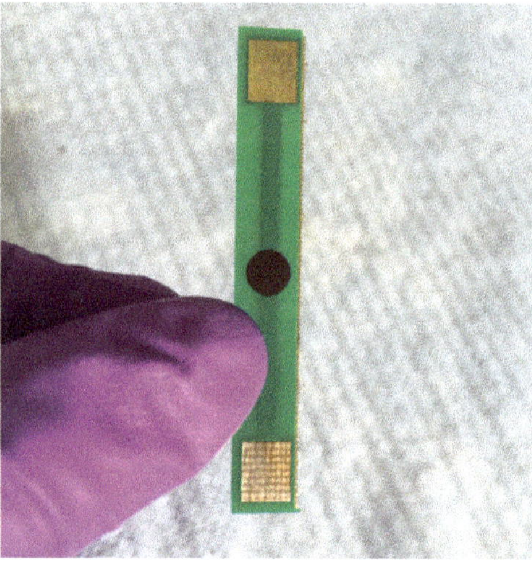

Figure 1. Photograph of a developed biosensor. The image displays the SWCNT-coated paper biosensor disc at the centre of the PCB substrate.

3.2. Biosensor Testing with BSA

Biosensors were tested with increasing concentrations of BSA standard solutions in addition to control solutions containing a non-target analyte. 5 μL of each solution was tested to generate a curve with increasing BSA concentration. After 5 μL of a particular solution was placed onto a biosensor disc, the reaction was allowed to proceed for 5 min before a measurement was recorded. This time-frame was chosen in order to allow for antibody-antigen complex formation in addition to the absorption of the sample on the disc.

With the goal of developing a biosensor based on the electrical percolation principle in order to facilitate label-free and simplified detection of protein biomarkers, we suspected that the magnitude of change in the electrical resistance of the developed biosensors following the addition of a sample solution would show a positive correlation with the concentration of BSA. A detailed schematic of the suspected sensing mechanism of our electrical percolation-based biosensor system is depicted in

Figure 2. As expected, our resulting curve shown in Figure 3 displays a positive correlation between BSA concentration and the magnitude of increased resistance. Based on these results, it is clear that the presence of the analyte, BSA, reduces the current through the biosensor, and hence, increases the resistance of the SWCNT-paper disc. Therefore, these results are in concordance with the mechanism of electrical percolation-based biosensors. In such biosensors, the formation of antibody-antigen complexes disrupts the continuity of a conductive network, causing an increase in the network's resistance, which our results exhibit [37].

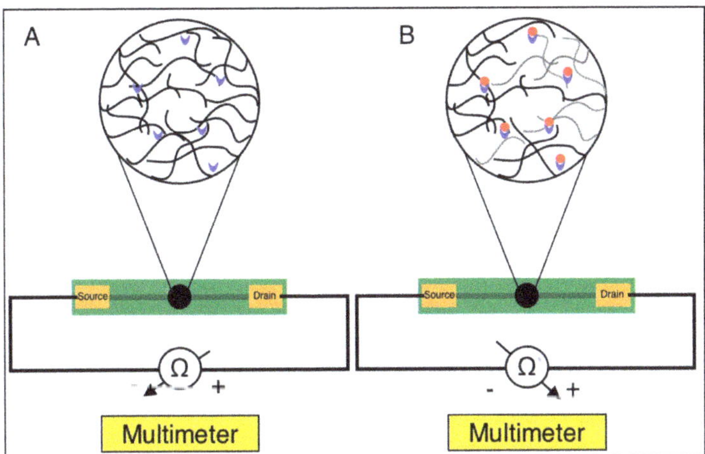

Figure 2. Sensing mechanism of the developed biosensors predicated on the electrical percolation principle with (**A**) The biosensor's SWCNT (**black**) and antibody (**blue**) network prior to the addition analyte and (**B**) the biosensor's SWCNT-antibody network following the addition of analyte (**red**), leading to an increase in resistance. Gray SWCNTs are those that are no longer connected in a conductive SWCNT pathway due to the presence of analyte.

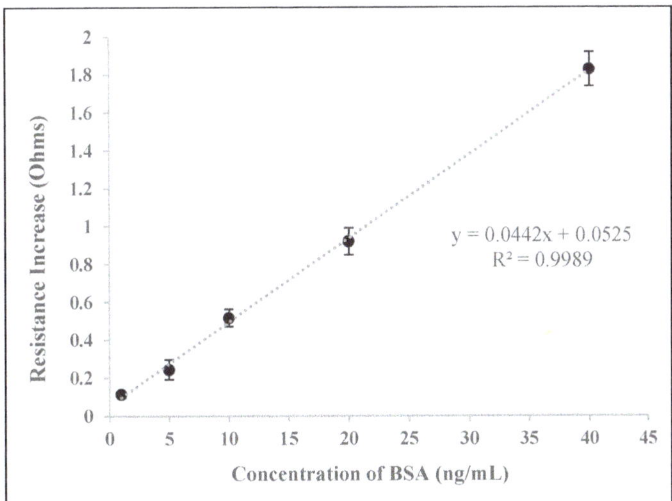

Figure 3. Biosensor detection results. This standard curve resulted from testing the developed biosensors with varying concentrations of BSA.

3.3. Temperature Optimization

The performance of electrochemical biosensors is often strongly influenced by the environmental conditions, such as the temperature at which the reaction proceeds. Temperature can influence biosensor performance by affecting various parameters such as the antibody-antigen reaction and SWCNT electron transfer. Thus, temperature is a variable that required optimization prior to performance assessment and comparisons with standard techniques such as ELISA. Accordingly, biosensor performance was tested with standard sample solutions at 4 °C, 23 °C, 37 °C and 50 °C (Figure 4). All of these temperatures are regularly deployed for the analysis of immunoreactions except for 50 °C, which was arbitrarily chosen to display a broad coverage of the effects of temperature. Between the temperatures that were tested, biosensors showed the most optimal performance at 37 °C. Beginning at 4 °C, the efficiency of the biosensors increased with temperature until 37 °C which is typically the most efficient temperature for antibody-antigen reactions. At 50 °C, efficiency of the biosensors was minimal again due to the denaturation of the antibodies. As 37 °C was found to be the optimal temperature for biosensor performance, we show a standard curve of the testing results under such conditions in Figure 3.

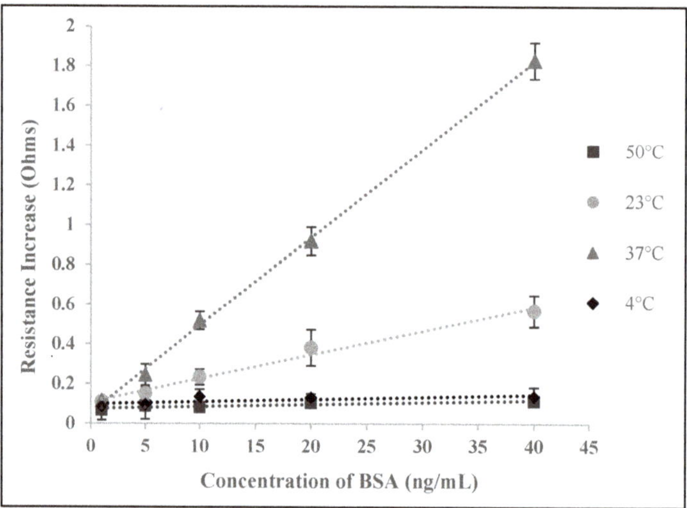

Figure 4. Temperature optimization of the sensing conditions for BSA. Biosensors were tested with various concentrations of BSA at 4 different temperatures.

3.4. Analysis of Biosensor Performance and Controlled Experiments

With the optimal temperature determined, we can evaluate overall biosensor performance and compare the developed biosensors to standard methods such as ELISA. At large, the performance of the developed biosensors greatly exceeded our expectations. With an LOD (limit of detection) of 2.89 ng/mL, the developed biosensors could sufficiently respond to even minor changes in BSA concentrations. The developed biosensors were also found to be highly specific as the addition of a non-target analyte solution, ovalbumin, and the addition of BSA to biosensors that were developed without BSA antibody resulted in only a slight change in resistance (Figure 5). Additionally, when increasing concentrations of ovalbumin were applied to the SWCNT-discs, or when increasing concentrations of BSA were applied to SWCNT-discs that were developed without the use of BSA antibody, there was no systematic correlation between the increasing concentrations and the magnitude of change in resistance. The slight changes in resistance were most likely a result of minor changes in the SWCNT network due to interference by the non-target or BSA analyte. Additionally, the weight of the solution may have caused strain in the SWCNT-coated disk or a modest change in morphology, leading to a slight increase in resistance. We suspect that with higher quality SWCNTs or further optimization of environmental conditions, the performance and LOD can be improved.

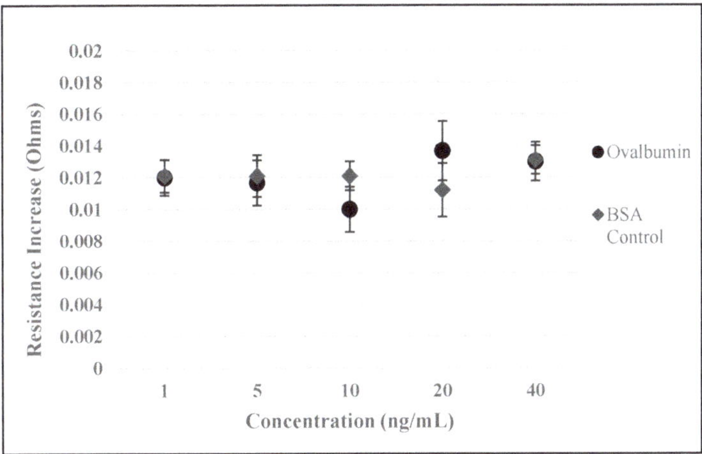

Figure 5. Control experiments with the developed biosensors. Developed biosensors were tested with various concentrations of a non-target analyte, ovalbumin. Biosensors were also developed without the use of BSA antibody and tested with various concentrations of BSA.

Our resulting standard curve for our developed biosensors displayed excellent linearity, with R^2 of 0.9989 (Figure 3). The curve is in the range of 0–40 ng/mL and even at 40 ng/mL, there is no deviation from linearity. We suspect that the biosensor's constituent antibodies are not fully saturated, and the biosensor may be capable of responding to more concentrated BSA solutions.

3.5. Analysis and Comparison with ELISA and Recently Developed Methods

We generated a standard curve with our prepared BSA concentrations using the traditional ELISA method in order to verify that our BSA test solutions were of the correct concentration (Figure 6). To generate this standard curve, we deployed a BSA ELISA kit that had an LOD of 6.26 ng/mL and a detection range of 6.25–400 ng/mL. As ELISA is a standard technique for the detection and measurement of proteins, both in the research lab and the clinical setting, it is of great interest to compare our developed biosensor system to the traditional ELISA method.

Our SWCNT-paper biosensor system's LOD of 2.89 ng/mL is comparable to the LOD of 6.25 ng/mL for the BSA ELISA kit. As our standard curve does not deviate from linearity over its entire range, we suspect that the SWCNT-paper biosensor's detection range has an upper limit that exceeds 40 ng/mL. In addition, our newly developed method takes no longer than 10 min while analysis by ELISA typically exceeds 4 h.

BSA was utilized in our experiments as a proof-of-concept, in order to determine whether the newly developed biosensor system could detect and measure a standard protein in solution. Overall, our results showed a positive correlation between BSA concentration and biosensor response. Thus, we suspect that our developed biosensors may be applied for the quantification of other proteins, simply by substituting specific antibodies. With the goal of developing a biosensor device that could potentially be deployed in the medical setting for the measurement of human proteins in bodily fluid, we used medically relevant (1–40 ng/mL) BSA concentrations for our experiments. For example, normal levels of prostate specific antigen (PSA), a prostate cancer biomarker in serum, are 0.5 to 2 ng/mL while the danger zone for PSA serum concentration is 4 to 10 ng/mL, a level that is indicative of the earliest stages of prostate cancer [38]. However, many biomarkers in human serum surpass this range, often reaching concentrations of 1–10 μg/mL [39].

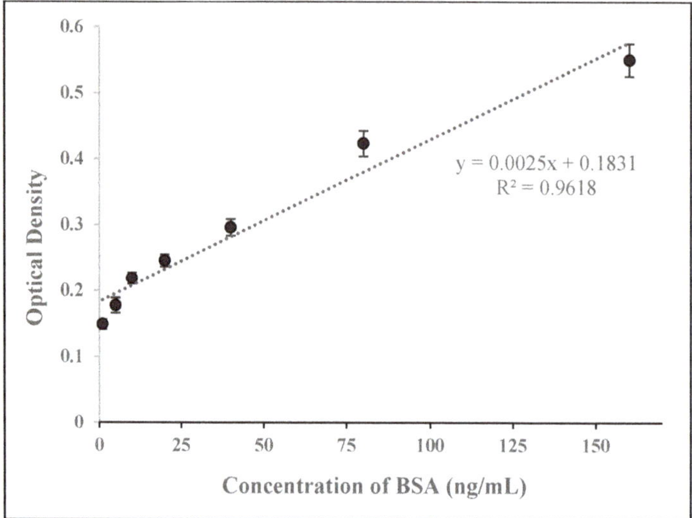

Figure 6. ELISA results for the prepared concentrations of BSA. The various concentrations of BSA that were used to analyze biosensor detection and performance were tested using ELISA for validation.

Thus, although an increased upper limit of detection for our presented methodology is suspected, further research should be carried out to determine whether it may be applicable for the measurement of protein biomarkers of a broadened range of concentrations.

Due to its medically relevant sensitivity, high specificity and rapid response time, our newly developed method may be more suitable for protein biomarker quantification than the traditional techniques often deployed in the clinical setting, such as ELISA. However, ELISA has a broad range of detection (6.25–400 ng/mL) for the quantification of BSA, and the presented method has only been tested with BSA concentrations within the 1–40 ng/mL range. Although the standard curve for the developed biosensors displayed a strong linear relationship between BSA concentration and biosensor response, and we appropriately suspect that the upper limit of detection may exceed 40 ng/mL, we cannot conclusively compare the detection range of the presented method with ELISA or other relevant methods until further research is carried out.

Additionally, the biosensor system is comprised of cellulose paper and utilizes minimal equipment, making the device more affordable than many standard techniques. The paper-based discs are disposable while the PCB substrate can be reused. In comparison to other rapid, sensitive and portable label-free biosensors for protein quantification that have recently been reported in the literature [16–28], the presented method has improved cost-effectiveness, simplification and ease of use.

4. Conclusions

We describe here an electrical percolation-based label free biosensor capable of quantifying a standard protein. With the goal of meeting the need for a rapid, simplified and affordable analytical device, the biosensors were developed using carbon nanotubes, a specific antibody and cellulose filtration paper. Based on the electrical-percolation principle, the biosensors were hypothesized to sense analytes on the basis of antibody-antigen complex formation, leading to a disruption of the SWCNT-antibody network's continuity and an increase in the network's resistance. The developed biosensors comprised an antibody specific to a standard protein, BSA, and were tested with various concentrations of this protein in order to evaluate our hypothesis. Our results show a positive correlation between BSA concentration and biosensor response. The detection limit of the newly developed biosensors was found to be comparable with ELISA. Additionally, the time necessary for detection was found to transcend that of ELISA, and the cost-effectiveness, ease of use and simplicity surpasses those of recent label-free biosensor developments. Further research needs to be carried out in order to conclusively compare the range of detection of the presented method with that of ELISA or other traditional methods. Importantly, the ability to substitute the specific antibody incorporated in the biosensor makes this method incredibly versatile. Due to the high sensitivity, specificity, affordability and rapidity of this device, we believe that it will find wide applications in both biological research and clinical diagnostics.

Supplementary Materials: The following are available online at http://www.mdpi.com/2079-6374/9/4/144/s1, Table S1: Source code for programmed Arduino Uno ohmmeter unit.

Author Contributions: Conceptualization, J.J.; methodology, J.J., M.L., P.W. and A.B.; Software, P.W.; validation, J.J.; formal analysis, J.J.; investigation, J.J.; resources, J.J., M.L. and A.B.; data curation, J.J.; writing-original draft preparation, J.J.; writing-review and editing, J.J. and A.B.; visualization, J.J.; supervision, A.B.; project administration, J.J.; funding acquisition, J.J.

Funding: This research received no external funding.

Acknowledgments: This work was performed at the Bioengineering for Biofuels and Bioproducts Lab and the Biomedical Instrumentation and Signal Analysis Lab at the University of Manitoba. Thank you to Moussavi, Lithgow and Levin for your help and guidance. Thank you to the graduate students and high school students working in both of these labs who offered support and assistance.

Conflicts of Interest: The authors declare no conflict of interest.

References

1. Unger-Saldana, K. Challenges to the early diagnosis and treatment of breast cancer in developing countries. *World J. Clin. Oncol.* **2014**, *5*, 465–477. [CrossRef] [PubMed]
2. Anderson, B.O.; Jakesz, R. Breast cancer issues in developing countries: An overview of the Breast Health Global Initiative. *World J. Surg.* **2008**, *32*, 2578–2585. [CrossRef] [PubMed]
3. Richards, M.A.; Westcombe, A.M.; Love, S.B.; Littlejohns, P.; Ramirez, A.J. Influence of delay on survival in patients with breast cancer: A systematic review. *Lancet* **1999**, *353*, 1119–1126. [CrossRef]
4. Rivera-Franco, M.M.; Rodriquez, E.L. Delays in breast cancer detection and treatment in developing countries. *Breast Cancer: Basic Clin. Res.* **2018**, *12*. [CrossRef]
5. Mabey, D.; Peeling, R.W.; Ustianowski, A.; Perkins, M.D. Diagnostics for the developing world. *Nat. Rev. Microbiol.* **2004**, *2*, 231–240. [CrossRef]
6. Singh, A.T.; Lantigua, D.; Meka, A.; Taing, S.; Pandher, M.; Camci-Unal, G. Paper-based sensors: Emerging themes and applications. *Sensors* **2018**, *18*, 2838. [CrossRef]
7. Luan, E.; Shoman, H.; Ratner, D.; Cheung, K.; Chrostowski, L. Silicon Photonic Biosensors Using Label-Free Detection. *Sensors* **2018**, *18*, 3519. [CrossRef]
8. Yamada, K.; Kim, C.; Kim, J.; Chung, J.; Lee, H.G.; Jun, S. Single walled carbon nanotube-based junction biosensor for detection of Escherichia coli. *PloS ONE* **2014**, *9*, e105767. [CrossRef]
9. Vestergaard, M.; Kerman, K.; Eiichi, T. An overview of label-free electrochemical protein sensors. *Sensors* **2007**, *7*, 3442–3458. [CrossRef]
10. Kerman, K.; Kobayashi, M.; Tamiya, E. Recent trends in electrochemical DNA biosensor technology. *Meas. Sci. Technol.* **2004**, *15*, R1. [CrossRef]
11. Wang, J. Electrochemical biosensors: Towards point-of-care cancer diagnostics. *Biosens. Bioelectron.* **2006**, *21*, 1887–1892. [CrossRef] [PubMed]
12. Kerman, K.; Vestergaard, M.; Nagatani, N.; Takamura, Y.; Tamiya, E. Electrochemical genosensor based on peptide nucleic acid-mediated PCR and asymmetric PCR techniques: Electrostatic interactions with a metal cation. *Anal. Chem.* **2006**, *8*, 2182–2189. [CrossRef]
13. Bini, A.; Minunni, M.; Tombelli, S.; Centi, S.; Mascini, M. Analytical performance of aptamer-based sensing for thrombin detection. *Anal. Chem.* **2007**, *79*, 3016–3019. [CrossRef] [PubMed]

14. Evgeni, E.; Cosnier, S.; Marks, R. Biosensors based on combined optical and electrochemical transduction for molecular diagnostics. *Expert Rev. Mol. Diagn.* **2011**, *11*, 533–546.
15. Sin, M.L.; Mach, K.E.; Wong, P.K.; Liao, J.C. Advances and challenges in biosensor-based diagnosis of infectious diseases. *Expert Rev. Mol. Diagn.* **2014**, *14*, 225–244. [CrossRef] [PubMed]
16. Mun, K.; Alvarez, S.D.; Choi, W.; Sailor, M. A stable, Label-free optical interferometric biosensor based on TIO_2 nanotube arrays. *ACS Nano* **2010**, *4*, 2070–2076. [CrossRef]
17. Wang, W.; Mai, Z.; Chen, Y.; Wang, J.; Li, L.; Su, Q.; Li, X.; Hong, X. A label-free fiber optic SPR biosensor for specific detection of C-reactive protein. *Sci. Rep.* **2017**, *7*, 16904. [CrossRef]
18. Chammem, H.; Hafaid, I.; Meilhac, O.; Menaa, F.; Mora, L.; Abdelghani, A. Surface plasmon resonance for C-reactive protein detection in human plasma. *J. Biomater. Nanobiotechnol.* **2014**, *5*, 153–158. [CrossRef]
19. Kim, Y.; Kim, J.P.; Han, S.J.; Sim, S.J. Aptamer biosensor for label-free detection of human immunoglobulin E based on surface plasmon resonance. *Sens. Actuators B. Chem.* **2009**, *139*, 471–475. [CrossRef]
20. Endo, T.; Kerman, K.; Nagatani, N.; Hiepa, H.M.; Kim, D.; Yonezawa, Y.; Nakano, K.; Tamiya, E. Multiple label-free detection of antigen-antibody reaction using localized surface plasmon resonance-based core-shell structures nanoparticle layer nanochip. *Anal. Chem.* **2006**, *78*, 6464–6475. [CrossRef]
21. Nascimento, N.M.; Juste-Dolz, A.; Grau-Garcia, E.; Roman-Ivorra, J.A.; Puchades, R.; Maquieira, A.; Morais, S.; Gimenez-Romero, D. Label-free piezoelectric biosensor for prognosis and diagnosis of Systemic Lupus Erythematosus. *Biosens. Bioelectron.* **2017**, *15*, 166–173. [CrossRef] [PubMed]
22. Lee, J.H.; Yoon, K.H.; Hwang, K.S.; Park, J.; Ahn, S.; Kim, T.S. Label free novel electrical detection using micromachined PZT monolithic thin film cantilever for the detection of C-reactive protein. *Biosens. Bioelectron.* **2004**, *20*, 269–275. [CrossRef] [PubMed]
23. Lee, J.H.; Hwang, K.S.; Park, J.; Yoon, K.H.; Yoon, D.S.; Kim, T.S. Immunoassay of prostate-specific antigen (PSA) using resonant frequency shift of piezoelectric nanomechanical microcantilever. *Biosens. Bioelectron.* **2005**, *20*, 2157–2162. [CrossRef] [PubMed]
24. Capobianco, J.A.; Shi, W.Y.; Adams, G.P.; Shih, W. Label-free Her2 detection and dissociation constant assessment in diluted human serum using a longitudinal extension mode of a piezoelectric microcantilever sensor. *Sens. Actuators B. Chem.* **2011**, *160*, 349–356. [CrossRef] [PubMed]
25. Lee, J.A.; Hwang, S.; Kwak, J.; Park, S.; Lee, S.S.; Lee, K. An electrochemical impedance biosensor with aptamer-modified pyrolyzed carbon electrode for label-free protein detection. *Sens. Actuators B. Chem.* **2008**, *129*, 372–379. [CrossRef]
26. Du, Y.; Li, B.; Wei, H.; Wang, Y.; Wang, E. Multifunctional label-free electrochemical biosensor based on an integrated aptamer. *Anal. Chem.* **2008**, *80*, 5110–5117. [CrossRef]
27. Ohno, Y.; Maehashi, K.; Matsumoto, K. Label-free biosensors based on aptamer-modified graphene field-effect transistors. *J. Am. Chem. Soc.* **2010**, *132*, 18012–18013. [CrossRef]
28. Wang, H.; Zhang, H.; Xu, S.; Gan, T.; Huang, K.; Liu, Y. A sensitive and label-free electrochemical impedance biosensor for protein detection based on terminal protection of small molecule-linked DNA. *Sens. Actuators B. Chem.* **2014**, *194*, 478–483. [CrossRef]
29. Thangamuthu, M.; Santschi, C.; Martin, O.J.F. Label-free electrochemical immunoassay for C-reactive protein. *Biosensors* **2018**, *8*, 34. [CrossRef]
30. Harrad, E.; Bourais, I.; Mohammadi, H.; Amine, A. Recent advances in electrochemical biosensors based on enzyme inhibition for clinical and pharmaceutical applications. *Sensors* **2018**, *18*, 164. [CrossRef]
31. Newman, J.D.; Turner, A.P. Home blood glucose biosensors: A commercial perspective. *Biosens. Bioelectron.* **2005**, *20*, 2435–2453. [CrossRef] [PubMed]
32. Tilmaciu, C.M.; Morris, M.C. Carbon nanotube biosensors. *Front. Chem.* **2015**, *3*, 59. [CrossRef] [PubMed]
33. Bruck, H.A.; Yang, M.; Kostov, Y.; Rasooly, A. Electrical percolation-based biosensors. *Methods* **2013**, *63*, 282–289. [CrossRef] [PubMed]
34. Charlier, J.C. Electronic and transport properties of nanotubes. *Rev. Mod. Phys.* **2007**, *79*, 677. [CrossRef]
35. Batra, B.; Pundir, C.S. An amperometric glutamate biosensor based on immobilization of glutamate oxidase onto carboxylated multiwalled carbon nanotubes/gold nanoparticles/chitosan composite film modified Au electrode. *Biosens. Bioelectron.* **2013**, *47*, 496–501. [CrossRef] [PubMed]
36. Chen, C.C.; Chou, Y.C. Electrical-conductivity fluctuations near the percolation threshold. *Phys. Rev. Lett.* **1985**, *54*, 2529–2532. [CrossRef] [PubMed]

37. Balasubramania, K.; Burghard, M. Biosensors based on carbon nanotubes. *Anal. Bioanal. Chem.* **2006**, *385*, 452–468. [CrossRef]
38. Rusling, J.F.; Kumar, C.V.; Gutkind, J.S.; Patel, V. Measurement of biomarker proteins for point-of-care early detection and monitoring of cancer. *Analyst* **2010**, *135*, 2496–2511. [CrossRef]
39. Kuang, Z.; Huang, R.; Yang, Z.; Lv, Z.; Chen, X.; Xu, F.; Yi, Y.; Wu, J.; Huang, R. Quantitative screening of serum protein biomarkers by reverse phase protein arrays. *Oncotarget* **2018**, *99*, 32624–32641. [CrossRef]

Article

Wireless Direct Microampere Current in Wound Healing: Clinical and Immunohistological Data from Two Single Case Reports

George Lagoumintzis [1,*], Zoi Zagoriti [1], Mogens S. Jensen [2], Theodoros Argyrakos [3], Constantinos Koutsojannis [4] and Konstantinos Poulas [1,*]

1 Department of Pharmacy, Laboratory of Molecular Biology & Immunology, University of Patras, 26504 Rio, Greece; zoizag@upatras.gr

2 Institute of Physic, Danish Technical University, DK-2800 Kgs Lyngby, Denmark; Mogens.Stibius.Jensen@fysik.dtu.dk

3 Department of Pathology, Evaggelismos General Hospital, GR 10676 Athens, Greece; lohengrin_e@yahoo.gr

4 Department of Physiotherapy, Laboratory of Health Physics and Computational Intelligence, University of Patras, 25100 Aigion, Greece; ckoutsog@teiwest.gr

* Correspondence: glagoum@upatras.gr (G.L.); kpoulas@upatras.gr (K.P.); Tel.: +30-2610-969953 (K.P.)

Received: 17 July 2019; Accepted: 2 September 2019; Published: 5 September 2019

Abstract: Chronic pressure ulcers are hard-to-heal wounds that decrease the patient's quality of life. Wireless Micro Current Stimulation (WMCS) is an innovative, non-invasive, similar to electrode-based electrostimulation (ES) technology, that generates and transfers ions that are negatively-charged to the injured tissue, using accessible air gases as a transfer medium. WMCS is capable of generating similar tissue potentials, as electrode-based ES, for injured tissue. Here, through immunohistochemistry, we intended to characterize the induced tissue healing biological mechanisms that occur during WMCS therapy. Two single cases of bedridden due to serious stroke white men with chronic non-healing pressure ulcers have been treated with WMCS technology. WMCS suppresses inflammatory responses by decreasing the aggregation of granulocytes, followed by stimulating myofibroblastic activity and a new formation of collagen fibers, as depicted by immunohistochemistry. As a result, WMCS provides a special adjunct or stand-alone therapy choice for chronic and non-healing injuries, similar to electrode-based ES, but with added (i.e., contactless) benefits towards its establishment as a routine clinical wound healing regime.

Keywords: chronic wounds; electrical stimulation; direct microcurrent; non-invasive; pressure ulcer; wireless technology

---

1. Introduction

The majority of chronic (i.e., non-healing) wounds and ulcers are associated with lifestyle-related diseases such as diabetes, cardiovascular pathologies, venous stasis disease, cancer, as well as with complications caused after the treatment of life-threatening diseases, such as stroke and cerebral palsy [1,2]. Chronic wounds affect more than 10 million people worldwide, and result in enormous health care expenditures, with the total cost estimated at more than 20 billion dollars per year [3–7]. Chronic wounds remain a challenging problem for every clinician, with complications contributing substantially to the rates of morbidity and mortality [8]. Understanding the physiology of healing and wound care with emphasis upon new therapeutic approaches is a prerequisite for acute and long-term wound management [4].

The wound healing process consists of four major phases: (i) Hemostasis, (ii) inflammation, (iii) proliferation and (iv) tissue remodeling [9]. These phases and their biochemical outcomes must occur in the proper sequence, at a specific time, and continue for a specific duration at an optimal intensity [2].

Wounds that exhibit impaired healing, including delayed acute wounds and chronic wounds, generally have failed to progress through the normal stages of healing. Such wounds frequently enter a state of pathologic inflammation due to a postponed, incomplete, or uncoordinated healing process [10]. There are many factors that can affect wound healing which interfere with one or more phases in this process, causing improper or impaired tissue repair [2].

As a consequence, the task of proper diagnosis and the treatment of difficult-to-heal wounds for most health professionals is not a straightforward process. Indeed, health care professionals are often challenged with perplex medical conditions in the field of chronic wound healing, despite the technological innovations and the emergence of a wide range of treatments for wounds.

Besides surgical treatment, various non-surgical approaches have been developed, and numerous drugs have been introduced to help chronic wounds. These include treatments such as bandages, vacuum-assisted closure, hyperbaric oxygen and maggot debridement therapy [11]. Alternatively, radiant heat dressing, ultrasound therapy, laser treatment, hydrotherapy, electromagnetic therapy and electrotherapy are other non-surgical approaches that have a scientific basis, and have therefore been advocated in the treatment of chronic wounds [12].

In this context, there is a considerable growing body of evidence on the indication and positive effects of electrical stimulation (ES) in wound healing [13–17]. ES has an extensive history, with records of its use since the 17th Century [18]. ES has been shown to be effective in accelerating wound repair using different currents and types of stimulation, thus it is currently recommended as an adjunct treatment for chronic ulcers and for reinitiating or accelerating the healing process of wounds [19,20]. In fact, previous studies have shown that alternating currents (AC) are useful in managing diabetic foot ulcers [21], and direct current (DC) has shown beneficial effects in the treatment of chronic skin ulcers [19]. Furthermore, pulsed current has been reported to enhance the healing of chronic wounds [22]. High-voltage ES has shown significant therapeutic results in healing chronic ulcers by increasing blood flow and oxygen concentration around the wound, and by directing cell migration and other components of the extracellular matrix [16,17]. The usual principle for electrode-based ES implementation is to transfer the current through moist surface electrode pads that are in electrolytic contact with both the external skin surface and the wound bed. Despite the considerable evidence on the beneficial effects of electrode-based ES in wound healing, it has not been widely adopted, mostly due to pain discomfort and the increased risk of infection upon the position of the electrodes next to the wound [16].

Recently, we and others have shown the clinical beneficial effects of an innovative technology, alternative to electrode-based ES, named Wireless Micro Current Stimulation (WMCS), which belongs to the non-invasive or non-contact treatment modalities of ES [20,22]. WMCS utilizes the current-carrying capacity of charged air gas, based on the ability of nitrogen (N_2) and/or oxygen (O_2) molecules to accept or donate electrons, in order to distribute currents and voltages within the tissue, comparable with the electrode-based ES method (see Supplementary Materials: Calculation Model S1).

Based on that specific ability of WMCS to generate microcurrent in the tissues in a comparable manner with electrode-based ES, here we extend our previous clinical applications and findings by reporting the results of treatments of two different types of wounds, using the WMCS method. We also present important experimental evidence on the improvement of these two representative cases with chronic, non-healing, pressure ulcers of different etiology and severity. Particularly, by a series of photographs, we monitor the course of wound healing over treatment with WMCS and we record the beneficial clinical outcome of the treated cases. Furthermore, by an immunohistochemical analysis of patient's biopsies we try to delineate the activation of the most characteristic subsequent cellular and molecular events that take place during the process of patients' wound healing.

2. Materials and Methods

2.1. Ethical Considerations

Written informed consent was obtained from all of the legally authorized guardians of patients to participate and to share their anonymized results for evaluation and publication. Tissue samples were taken for immunohistochemical analysis. Patients could withdraw for any reason upon their guardians' decision, without risk to their ongoing standard of care. The study was conducted in accordance with the Declaration of Helsinki, and the protocol was approved by the Ethics Committee of the University of Patras (Patras, Greece) (67257/2019).

2.2. Study Participants—Cases Presentation

Patients exhibited chronic, non-healing pressure ulcer wounds, persisted for ~4–6 weeks. The patients were followed clinically for adverse health events, including infection, before and after the biopsies were performed and during the Wireless Micro Current Stimulation (WMCS) treatment. Standard serial photography was used to record the wounds periodically. The detailed medical history of the two patients participating in this study is described below:

Patient 1: A 62-year-old white male, bedridden and unconscious for the last year after a severe stroke, presented with a great pressure ulcer at his coccyx (~11 cm × 24 cm × 8 cm deep). The ulcer arose probably due to his prolonged bed rest. Primarily, conventional wound care was adopted for the healing process (i.e., washing, disinfection, topical dressing, etc.), but the size of the ulcer gradually expanded. No history of diabetes was reported. Artificial nutrition was administered to the patient. The ulcer persisted, and it was gradually expanding for at least six weeks prior to initiation with the WMCS device. Treatment with the WMCS was initiated, using 1.5 μA current for 45 min, three times per week, for a total of ~9 months. Standard wound care by cleaning with normal saline was routine before the WMCS therapy besides its daily conventional wound care. Necrotic tissue was removed completely as well as fibrin or other coverings if necessary. During the 9-month course of treatment, systemic antibiotics were administered twice to the patient due to a common cold illness; WMCS therapy was then also paused for about a week, then continued as described above. No pain medication was administered to the patient.

Patient 2: A 69-year-old white male patient with a hard-to-heal bed sore ulcer on his left heel (~2 cm × 2 cm and 0.5 cm deep) participated in this study. The patient experienced a stroke three months ago, and he was gradually recovering. However, his mobility was impaired; thus, he was lying upon his bed while recovering. The ulcer appeared due to this prolonged bed rest. Conventional wound dressing was initially adopted for the healing process, but the size of the ulcer remained constant. The wound persisted—while gradually slightly expanding—for about four weeks prior initiation with the WMCS technology. Treatment with the WMCS was initiated, using 1.5 μA current for 45min daily for a total of 15 days. Standard wound care was routine before the WMCS therapy, while no topical antibacterial agent, pain medication or systemic antibiotics were administered during the course of the WMCS treatment.

2.3. WMCS Device and Set-Up

The WMCS device used in this study is a prototype based on a commercially-available model (W200, Wetling, Fredensborg, Denmark), capable of generating tissue currents and voltages as a conventional electrode-based electrostimulation (ES) device. WMCS utilizes the current-carrying capacity of charged air gas, based on the ability of O_2 to accept or donate electrons, respectively, thus "spraying" airborne O_2^- to the skin with the aid of an accelerator equipment. The device used to produce the O_2^--induced current in the patient's body is shown diagrammatically in Scheme 1, in a typical exposure condition. The flow of O_2^- ions from the device is directed towards the target (i.e., the patient), who is isolated from the ground. The device is capable of producing a specific number of charged particles which are covering the area to be treated and by this way a micro-current of 1.5–4.0 μA intensity is generated.

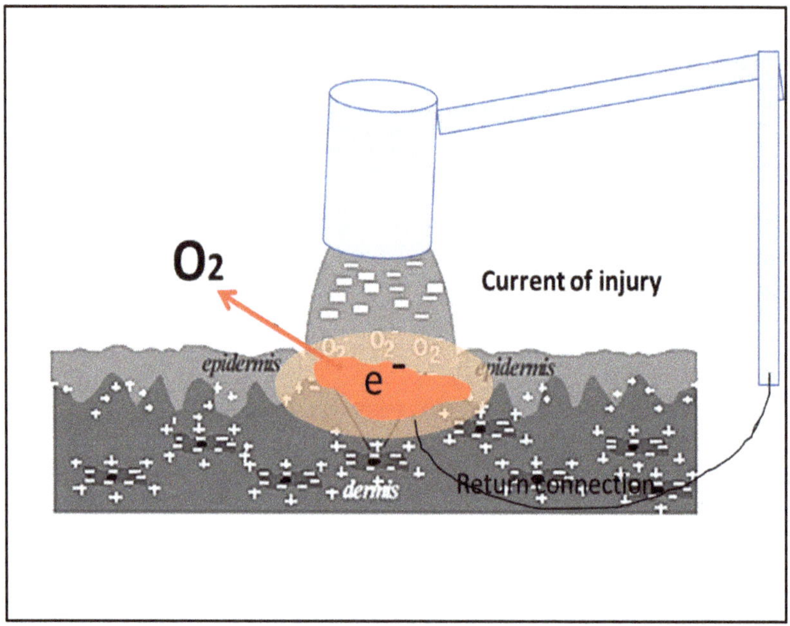

Scheme 1. Schematic representation of the Wireless Micro Current Stimulation (WMCS) device, in a typical "spraying" of generated airborne O_2^- onto a wound.

The patient, while lying in bed, is connected to the WMCS device through a neutral electrode. The return path, established by wrist or ankle straps, has a dual purpose. It makes the measurement of exposure possible, and it maintains the target at virtually zero potential with respect to the ground. A control box permits the adjustment of the current and treatment duration. The intensity was set to 1.5 μA for all of the therapies, and the distance from the head of the device to any wound was 12–15 cm. If the head of the device was too far from the wound area for the transferred electric effect to reach the patient, there was a continuous warning tone.

2.4. Immunohistochemical Analysis of Biopsies

Standard—but not identical—immunohistochemical analysis was performed on each patient's specimens. In particular, Hematoxylin and Eosin stain was done to visualize any signs of inflammation, Masson's trichrome stain in order to clearly differentiate the collagen fibers and c-kit antigen as a marker of mast cell existence. Both Patients' wound biopsies were taken by a qualified general practitioner at indicative time-points during the WMCS treatment. Both patients underwent a full-thickness wedge biopsy of the wound's edge. Samples were stored in paraffin blocks, and then 3 μm histological sections were made using a histological microtome. The specimens were analyzed at the Department of Pathology, Evaggelismos Hospital, Athens, Greece, by routine semi-quantitative immunohistochemistry, as described below, and observed under a Leica DM2000 microscope. Cellular and/or protein findings were descriptively quantified without the aid of any computerized resources, by a single pathologist, in order to avoid measurement biases by multiple measurers.

2.5. Hematoxylin and Eosin

Slides were placed in staining jar and deparaffinized by submerging into three series of absolute xylene for four minutes, followed by 100%, 95%, 90% and 70% of ethanol for four minutes of each percentage. These slides were then washed in running tap water for two minutes.

After this, slides were submerged into Harris Hematoxylin for two minutes and then washed in running tap water for another two minutes. The slides were then submerged into 1% acid alcohol for three dips to decolorize it, and washed in running tap water for two minutes. Next, the slides were submerged into 2% potassium acetate for three minutes and again washed in running tap water for two minutes. Slides were then submerged into hematoxylin for 10 min. After that, slides were submerged into Eosin for two minutes, followed by washing in running tap water for a further two minutes. Before observation, slides were dipped into absolute xylene for one minute and finally mounted with cover slip using DPX mounting.

2.6. Human CD117/c-Kit 17 Antigen

Immunohistochemistry was performed on formalin-fixed and paraffin-embedded 3 μm thick sections. Tissue sections were deparaffinized in xylene and rehydrated in decreasing ethanol series. For antigen retrieval, sections were boiled in 0.01 M citrate buffer (pH = 6, slightly acidic) for 10 min. Methanol with 0.5% H_2O_2 was used to block endogenous peroxidase activity for 10 min. Tissue sections were then washed in Tris-buffered saline (TBS, pH = 7.6, slightly alkali) and incubated with diluted normal serum for 10 min and then treated with primary antibody for 30 min, according to the procedure outlined by the manufacturer. The applied primary antibody was c-kit monoclonal antibody (Novocastra RTU-CD117, RE 7290 k, Newcastle, UK). After rinsing with TBS, the sections were incubated with secondary antibody (Dako Envision Dual Link Labeled Polymer HRP, Glostrup, Denmark), washed again in TBS and reacted with diaminobenzine hydrochloride (DAB) for five minutes. Finally, slides were counterstained with hematoxylin and cover-slipped with a synthetic mounting media.

2.7. Masson's Trichrome

For visualization of collagen fibers and the histological assessment of collagen deposition, trichrome staining was performed using the Masson Trichrome Staining Kit (Sigma Aldrich, St. Louis, MO, USA). First, the slides were deparaffinized and rehydrated through a descending manner of alcohol: 100% alcohol, 95% alcohol and 70% alcohol, for four minutes in each percentage. The slides were washed in distilled water. The slides then were submerged in warmed Bouin's solution at 60 °C for 45 min. Next, the slides were washed in running tap water until the yellow color in samples disappeared. To differentiate nuclei, slides then were immersed in modified Wiegert's hematoxylin for eight minutes, after that washed in running water for two minutes. In order to stain the cytoplasms and erythrocytes, the slides were submerged in anionic dyes, acid fuschin for five minutes; then again these slides were washed with running tap water for two minutes. Next, slides were treated with a phosphomolybidic acid solution for another 10 min as a mordant, and immediately the slides were submerged into methyl blue solution for five minutes in order to stain fibroblast and collagen. After that, slides were washed in running water for two minutes and lastly treated with 1% acetic acid solution for one minute. Slides then were dehydrated into a series of alcohol of 70%, 80%, 95% and 100%, respectively, for one minute within each percentage. Before observation, slides were dipped into absolute xylene for one minute and finally mounted with cover slip using DPX mounting. Finally, the slides were examined under the microscope by the same pathologist blinded to the treatments.

2.8. Measurement of Wound Closure

Wound closure was measured by serial standard photographs taken at several time intervals during the WMCS treatment. The target wound's greatest length and width were measured at baseline.

The ulcer size was calculated by multiplying the length and width. Percentage wound area reduction was calculated by dividing the wound area at the indicative time point after WMCS treatment with the wound area before WMCS treatment (Day = 0), and then multiplied by 100. The entire process of area measurement was done by a single person who was blinded to the group of the patient and the session for each wound being measured.

3. Results

3.1. Clinical Outcome

The clinical picture of Patient One's massive pressure ulcer prior WMCS treatment is shown in Figure 1A (a–i). A clear change in wound edges with an obvious reduction in the wound's dimensions and total area was observed after a period of ~3-month treatment Figure 1A (c). A 50% of wound closure was achieved after a 6-month period of treatment Figure 1A (d–f), while a more than 90% wound closure and regrowth of normal skin was seen by the end of a total of 8 to 9 month WMCS treatment Figure 1A (g–i).

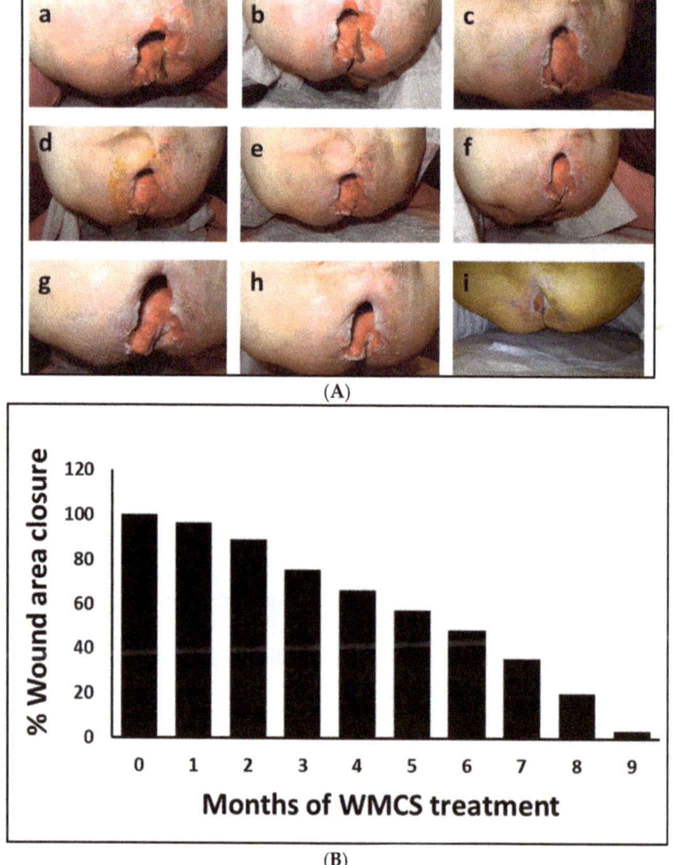

Figure 1. (**A**) (a–i): Serial photography of the clinical outcome of Patient 1 after WMCS treatment. (**B**) Presentation of photographic wound closure percentage of Patient 1. Wound healing is expressed as percentage wound area closure relative to original size [(wound area)/(original wound area)].

Measurement of the wounded area is one of the key aspects in the assessment of the healing process, since it can indicate healing improvement. Figure 1B shows the percentage (%) of wound closure for Patient 1 from photographs taken immediately after treatment with WMCS. As shown, wound healing started gradually after WMCS treatment, with an obvious macroscopic improvement even at the end of the 1st month of treatment. At approximately six months of treatment, a 50% wound closure was achieved, while by the end of the 9-month treatment a greater than 95% of wound closure was observed as compared to the initial wound area before WMCS treatment.

For Patient 2, the clinical picture of the wound prior WMCS treatment is shown in Figure 2A (a). After three sessions, healthy new epithelial tissue was visible in the wound bed and at the wound edge, Figure 2A (b). A clear reduction in the wound's dimensions and total area was achieved after 6 sessions, Figure 2A (c), allowing for no dressing changes. Continued decrease in scar tissue and regrowth of normal skin were observed from day 9 to day 15, Figure 2A (d–f), with improved wound contraction and significant re-epithelization.

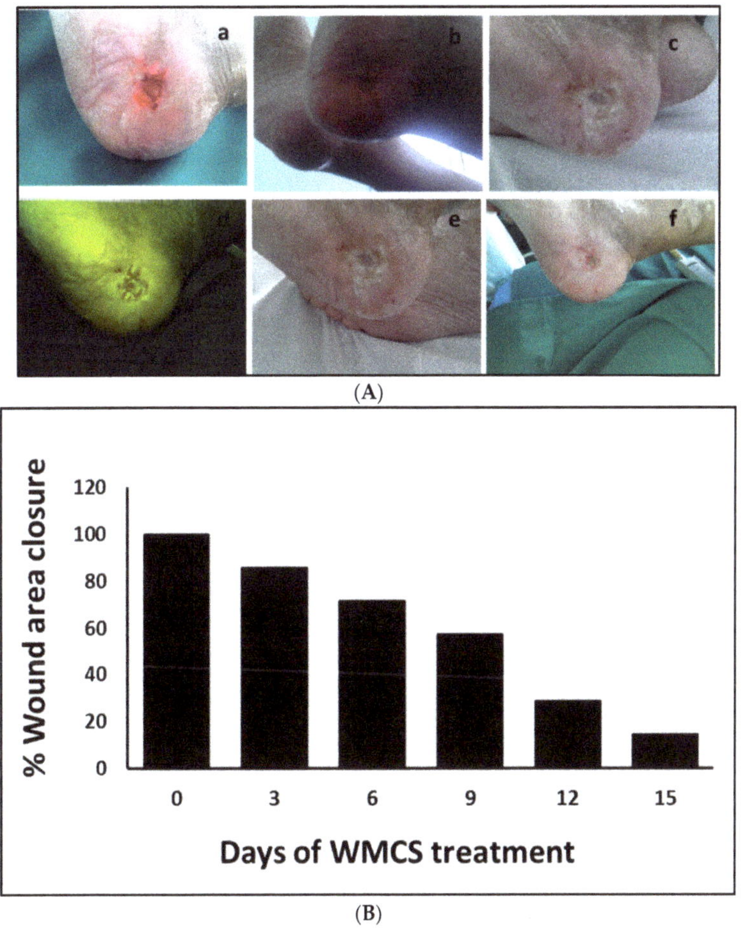

Figure 2. (**A**) (a–f): Serial photography of the clinical outcome of Patient 2 after WMCS treatment. (**B**) Presentation of the photographic wound closure percentage of Patient 2. Wound healing is expressed as percentage wound area closure relative to original size [(wound area)/(original wound area)].

Figure 2B is showing the percentage (%) of wound closure for Patient 2 as a factor of time treatment with the WMCS technology from photographs taken after treatment. As we can see, this wound started gradually to heal immediately after WMCS treatment, with more obvious changes in wound dimensions after a week (Day 6 to 7). An approximately 50% of wound area reduction in size was observed at day 8 to 9, whereas at day 15 with WMCS treatment, the wound was almost entirely healed.

3.2. Immunohistochemical Analysis

Monitoring healing progress by tissue biopsies of the wounded area is the hallmark of the evaluation of an active and proper wound healing process. Thus, tissue biopsies were immunohistochemically analyzed in order to identify potential cellular events upon WMCS treatment that are indicative of a restart of wound healing. Figure 3A–C shows the Hematoxylin and Eosin stain of tissue biopsy from Patient 1. In detail, Figure 3A depicts signs of active granulocyte aggregation with some signs of inflammation after 1-month WMCS treatment sessions. A suppression of inflammation upon WMCS treatment (2-month) as evidenced followed by an increase in myofibroblastic activity and a reduced granulocyte aggregation, is demonstrated in Figure 3B. After 3-month of WMCS treatment (Figure 3C), microscopic examination revealed further established myofibroblastic activity and almost totally diminished granulocyte aggregation.

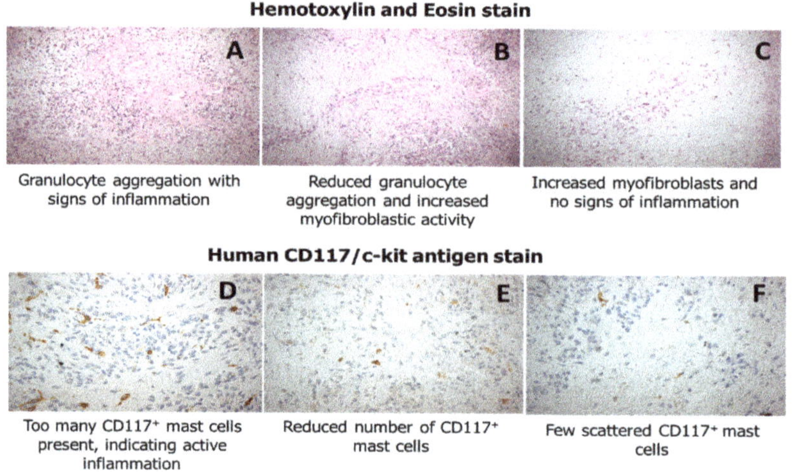

Figure 3. (**A–C**) Hemotoxylin and Eosin immunohistochemistry of tissue biopsies from Patient 1, demonstrating initial granulation tissue formation and inflammation markers and increased myofibroblastic activity, and (**D–F**) CD117+ mast cells after WMCS treatment.

Increased numbers of mast cells (c-kit antigen stain, panels D–F) are present at early stages of wound healing, indicating an active process of inflammatory reactions (Figure 3D). WMCS treatment seems to affect the activation of mast cells, as an increased number of visible mast cells is observed after 1-month WMCS treatment (Figure 3D), ultimately leading to the observation of few scattered mast cells between 2-month and 3-month of WMCS therapeutic treatments (Figure 3E,F).

Tissue biopsies from Patient 2 were stained apart from H&E stain with the Masson's trichrome stain and the c-kit antigen for collagen formation, myofibroblastic proliferation and mast cell existence after WMCS treatment. As shown in Figure 4A–C, a gradually increased amount of collagen deposition during the healing process is observed after WMCS treatment.

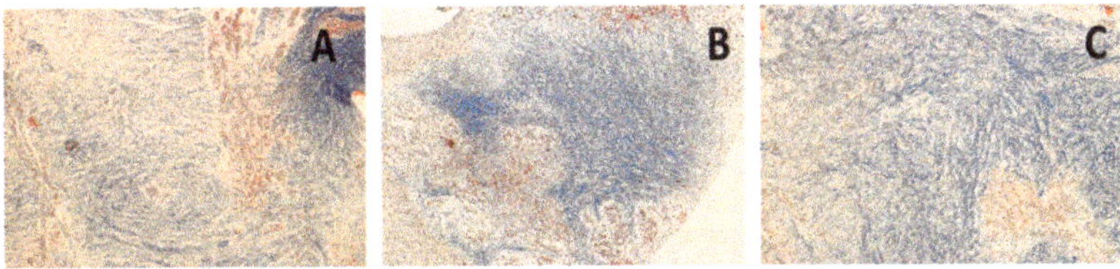

Figure 4. (**A–C**) Masson's trichrome stain of tissue biopsies from Patient 2. Masson's stain indicating collagen formation and myofibroblastic proliferation after WMCS treatment.

Moreover, WMCS enhances myofibroblastic proliferation (Figure 5A), yet with a relative focal increase of mast cells after three Days (Figure 5B). Finally, after six Days of WMCS treatment few scattered mast cells are seen as revealed by the c-kit antigen (Figure 5C).

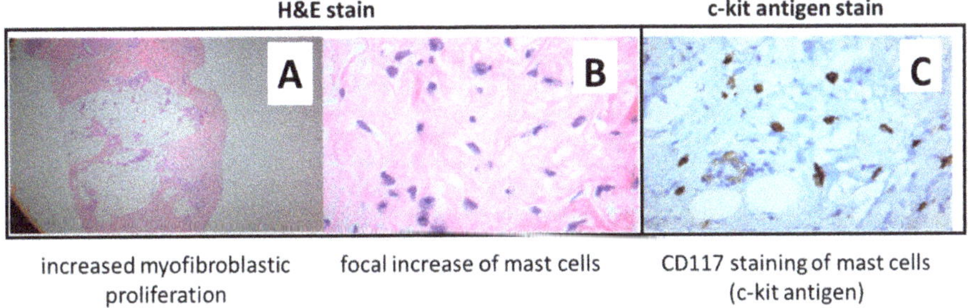

Figure 5. (**A–C**) Immunohistochemical analysis of tissue biopsies from Patient 2, showing increased myofibroblastic activity, with a relative focal increase of mast cells (H&E stain 5A and 5B, respectively), and c-kit (CD117) antigen stain showing reduced CD117+ mast cells over WMCS treatment (5C).

4. Discussion

Pressure ulcers, also called bedsores or decubitus ulcers, develop when there is too much unrelieved pressure on the skin in combination with shear and/or friction. They form typically over a bony prominence, leading to ischemia, cell death and tissue necrosis [23,24]. They are more common in bedridden patients, affecting their quality of life, morbidity and mortality. Once a pressure ulcer develops, not only does it cause pain and discomfort, but it may lead to complications such as infection, like meningitis, cellulitis and endocarditis, with the potential for sepsis and death [25]. The shoulder blades, tailbone, elbows, heels and hips are the most common sites for bed sores because these areas contain little muscle and fat [26]. The development of a pressure ulcer can interfere with functional recovery, promote social isolation, and contribute to excessive length of hospital stay [27]. As a consequence, they have been described as one of the costliest and physically devastating complications [28,29], and they are the third most expensive disorder after cancer and cardiovascular diseases [30].

Treatment of pressure ulcers involves multiple methods intended to alleviate the conditions contributing to ulcer development (support surfaces, repositioning and nutritional support), protection of the wound from contamination, and the creation of a clean wound environment, promotion of tissue healing (local wound applications, debridement and wound cleansing), adjunctive therapies, as well as consideration for surgical treatment [31]. Understanding the function of bioelectricity in wound healing offers a possibility for the therapeutic implementation of ES, especially when natural regeneration mechanisms have been paused. ES is a well-proven technique for the therapy of chronic wounds [32]. Its application for the treatment of wounds adopts the physiological fact that skin and epithelia act as batteries and have electric potentials.

When a wound occurs in the epithelium, there is an electrical leak that short-circuits the skin, enabling the current to flow out of the wound, resulting in the current of injury. This leak of current provides signals for the migration of the epithelia into the wound, initiating a physiological endogenous regeneration processes [32–34]. Disruption of these physiological electric fields and ionic currents alters normal organ development, tissue regeneration, and thus, proper wound healing [35–37].

Our team has previously experimented on WMCS, an innovative modality of traditional electrode-based ES, and we have particularly shown recently that WMCS can be an effective method of diabetic-related wound care [22], for difficult to treat chronic wounds [20], for firework burns [38], as well as for Martorell's ulcer [39]. Tissue potentials and currents generated by either electrode-based ES or WMCS seem to be the determinant factor which initially drives the beneficial clinical outcomes of these two similar, but distinct, techniques.

In this report, based upon clinical as well as on immunohistochemical evidences, we further report on WMCS treatment with two single case reports of distinct pathologies. Our results showed that WMCS aids the wound healing of pressure ulcers by restoring the natural current of injury, thus, reinitiating the body's tissue physiological regeneration mechanisms. Notably, WMCS suppresses inflammatory reactions by reducing granulocyte aggregation, followed by stimulating myofibroblastic activity and collagen fiber formation. These results are in accordance, not only with our previous clinical observational studies, but also to others, who have effectively used similar ES modalities to treat pressure ulcers [40–42]. In addition, several recent systematic reviews provide positive recommendation regarding the effectiveness of ES to increase wound healing [43–48]. Consequently, based on literature as well as on our previous work, WMCS offers a unique treatment option to chronic and non-healing wounds of diverse etiopathology. WMCS in particular, is believed to aid in wound healing by imitating and restoring the natural electrical current that is been disrupted in injured skin.

5. Conclusions

WMCS has been both theoretically, clinically and experimentally, evidently proven as an effective, easy-to-use, non-invasive and time-efficient treatment for the wound healing of different etiopathologies, as electrode-based ES; however, with more of its advantages reflecting its contactless application. Yet, greater knowledge of the specific repairing mechanisms upon WMCS treatment will convey us toward more effective and broad clinical applications. To this end, more experimental and clinical efforts are needed to elucidate how wound regeneration is influenced by the choice of optimal parameters of WMCS (i.e., polarity, frequency, duration) that could affect the general healing process in each medical case.

Supplementary Materials: The following are available online at http://www.mdpi.com/2079-6374/9/3/107/s1, Calculations S1: Calculation Model of Tissue Potentials Between WMCS and (traditional) Electrode-Based ES Method.

Author Contributions: Conceptualization, G.L., K.P.; methodology, Z.Z., G.L., T.A., C.K.; data curation, K.P.; mathematical calculations, M.S.J.; writing—original draft preparation, G.L.; writing—review and editing, G.L., Z.Z.; supervision, G.L., K.P.

Funding: This research received no external funding.

Acknowledgments: We thank the patients for making this report possible. Authors are also grateful to the National Research Infrastructure Program of the University of Patras "OMIC-ENGINE-Synthetic Biology: from omics technologies to genomic engineering" for financial support of the personnel.

Conflicts of Interest: The authors declare no conflict of interest.

References

1. Avishai, E.; Yeghiazaryan, K.; Golubnitschaja, O. Impaired wound healing: Facts and hypotheses for multi-professional considerations in predictive, preventive and personalised medicine. *EPMA J.* **2017**, *8*, 23–33. [CrossRef] [PubMed]

2. Guo, S.; DiPietro, L.A. Factors affecting wound healing. *J. Dent. Res.* **2010**, *89*, 219–229. [CrossRef] [PubMed]
3. Augustin, M.; Brocatti, L.K.; Rustenbach, S.J.; Schäfer, I.; Herberger, K. Cost-of-illness of leg ulcers in the community. *Int. Wound, J.* **2014**, *11*, 283–292. [CrossRef] [PubMed]
4. Frykberg, R.G.; Banks, J. Challenges in the treatment of chronic wounds. *Adv. Wound Care* **2015**, *4*, 560–582. [CrossRef] [PubMed]
5. Järbrink, K.; Ni, G.; Sönnergren, H.; Schmidtchen, A.; Pang, C.; Bajpai, R.; Car, J. The humanistic and economic burden of chronic wounds: A protocol for a systematic review. *Syst. Rev.* **2017**, *6*, 15. [CrossRef] [PubMed]
6. Phillips, C.J.; Humphreys, I.; Fletcher, J.; Harding, K.; Chamberlain, G.; Macey, S. Estimating the costs associated with the management of patients with chronic wounds using linked routine data. *Int. Wound J.* **2016**, *13*, 1193–1197. [CrossRef] [PubMed]
7. Posnett, J.; Franks, P.J. The burden of chronic wounds in the UK. *Nurs. Times* **2008**, *104*, 44–45. [PubMed]
8. Agale, S.V. Chronic leg ulcers: Epidemiology, aetiopathogenesis, and management. *Ulcers* **2013**, *2013*, 1–9. [CrossRef]
9. Gosain, A.; DiPietro, L.A. Aging and wound healing. *World J. Surg.* **2004**, *28*, 321–326. [CrossRef]
10. Diegelmann, R.F.; Evans, M.C. Wound healing: An overview of acute, fibrotic and delayed healing. *Front. Biosci.* **2004**, *9*, 283–289. [CrossRef]
11. Han, G.; Ceilley, R. Chronic wound healing: A review of current management and treatments. *Adv. Ther.* **2017**, *34*, 599–610. [CrossRef] [PubMed]
12. Korzendorfer, H.; Hettrick, H. Biophysical technologies for management of wound bioburden. *Adv. Wound Care* **2014**, *3*, 733. [CrossRef] [PubMed]
13. Janssen, T.; Smit, C.; Hopman, M. Prevention and treatment of pressure ulcers using electrical stimulation. In *Pressure Ulcer Research*; Bader, D.L., Bouten, C., Colin, D., Oomens, C.W.J., Eds.; Springer-Verlag; Berlin/Heidelberg, Germany, 2005; pp. 89–107. ISBN 978-3-540-28804-6.
14. Liu, L.Q.; Moody, J.; Traynor, M.; Dyson, S.; Gall, A. A systematic review of electrical stimulation for pressure ulcer prevention and treatment in people with spinal cord injuries. *J. Spinal Cord Med.* **2014**, *37*, 703–718. [CrossRef] [PubMed]
15. Messerli, M.A.; Graham, D.M. Extracellular electrical fields direct wound healing and regeneration. *Biol. Bull.* **2011**, *221*, 79–92. [CrossRef] [PubMed]
16. Tai, G.; Tai, M.; Zhao, M. Electrically stimulated cell migration and its contribution to wound healing. *Burn. Trauma* **2018**, *6*, 20. [CrossRef] [PubMed]
17. Wang, E.; Zhao, M. Regulation of tissue repair and regeneration by electric fields. *Chin. J. Traumatol.* **2010**, *13*, 55–61. [PubMed]
18. Kambouris, M.; Zagoriti, Z.; Lagoumintzis, G.; Poulas, K. From therapeutic electrotherapy to electroceuticals: Formats, applications and prospects of electrostimulation. *Annu. Res. Rev. Biol.* **2014**, *4*, 3054–3070. [CrossRef]
19. Adunsky, A.; Ohry, A.; DDCT Group. Decubitus direct current treatment (DDCT) of pressure ulcers: Results of a randomized double-blinded placebo controlled study. *Arch. Gerontol. Geriatr.* **2005**, *41*, 261–269. [CrossRef]
20. Guerriero, F.; Botarelli, E.; Mele, G.; Polo, L.; Zoncu, D.; Renati, P.; Sgarlata, C.; Rollone, M.; Ricevuti, G.; Maurizi, N.; et al. Effectiveness of an innovative pulsed electromagnetic fields stimulation in healing of untreatable skin ulcers in the frail elderly: Two case reports. *Case Rep. Dermatol. Med.* **2015**, *2015*, 1–6. [CrossRef]
21. Lundeberg, T.C.; Eriksson, S.V.; Malm, M. Electrical nerve stimulation improves healing of diabetic ulcers. *Ann. Plast. Surg.* **1992**, *29*, 328–331. [CrossRef]
22. Castana, O.; Dimitrouli, A.; Argyrakos, T.; Theodorakopoulou, E.; Stampolidis, N.; Papadopoulos, E.; Pallantzas, A.; Stasinopoulos, I.; Poulas, K. Wireless electrical stimulation. *Int. J. Low. Extrem. Wounds* **2013**, *12*, 18–21. [CrossRef] [PubMed]
23. Ramadhinara, A.; Poulas, K. Use of wireless microcurrent stimulation for the treatment of diabetes-related wounds. Adv. *Skin Wound Care* **2013**, *26*, 1–4. [CrossRef] [PubMed]
24. Bansal, C.; Scott, R.; Stewart, D.; Cockerell, C.J. Decubitus ulcers: A review of the literature. *Int. J. Dermatol.* **2005**, *44*, 805–810. [CrossRef] [PubMed]

25. *Prevention and Treatment of Pressure Ulcers: Quick Reference Guide*; Cambridge Media: Cambridge, UK, 2009; ISBN 978-0-9579343-6-8.
26. Bolton, L. Evidence corner: Evidence-based care for malignant wounds. *Wounds Compend. Clin. Res. Pract.* **2016**, *28*, 214–216.
27. Moore, Z.E.; Webster, J.; Samuriwo, R. Wound-care teams for preventing and treating pressure ulcers. *Cochrane Database Syst. Rev.* **2015**, *16*, CD011011. [CrossRef] [PubMed]
28. Reddy, M.; Gill, S.S.; Rochon, P.A. Preventing pressure ulcers: A systematic review. *JAMA* **2006**, *296*, 974. [CrossRef] [PubMed]
29. Burdette-Taylor, S.R.; Kass, J. Heel ulcers in critical care units: A major pressure problem. *Crit. Care Nurs. Q.* **2002**, *25*, 41–53. [CrossRef] [PubMed]
30. Demarré, L.; Van Lancker, A.; Van Hecke, A.; Verhaeghe, S.; Grypdonck, M.; Lemey, J.; Annemans, L.; Beeckman, D. The cost of prevention and treatment of pressure ulcers: A systematic review. *Int. J. Nurs. Stud.* **2015**, *52*, 1754–1774. [CrossRef]
31. Pattanshetty, R.B.; Prasade, P.M.; Aradhana, K.M. Risk assessment of decubitus ulcers using four scales among patients admitted in medical and surgical intensive care units in a tertiary care set up: A cross-sectional study. *Int. J. Physiother. Res.* **2015**, *3*, 971–977. [CrossRef]
32. Reddy, M.; Gill, S.S.; Kalkar, S.R.; Wu, W.; Anderson, P.J.; Rochon, P.A. Treatment of pressure ulcers. *JAMA* **2008**, *300*, 2647. [CrossRef]
33. Enoch, S.; Price, P.E. Cellular, molecular and biochemical differences in the pathophysiology of healing between acute wounds, chronic wounds and wounds in the aged. *World Wide Wounds* **2004**, *2005*, 12.
34. Snyder, S.; DeJulius, C.; Willits, R.K. Electrical stimulation increases random migration of human dermal fibroblasts. *Ann. Biomed. Eng.* **2017**, *45*, 2049–2060. [CrossRef] [PubMed]
35. Piccolino, M. Luigi Galvani's path to animal electricity. *C. R. Biol.* **2006**, *329*, 303–318. [CrossRef] [PubMed]
36. Finger, S.; Piccolino, M.; Stahnisch, F.W. Alexander von Humboldt: Galvanism, animal electricity, and self-experimentation part 2: The electric eel, animal electricity, and later years. *J. Hist. Neurosci.* **2013**, *22*, 327–352. [CrossRef]
37. Foulds, I.S.; Barker, A.T. Human skin battery potentials and their possible role in wound healing. *Br. J. Dermatol.* **1983**, *109*, 515–522. [CrossRef] [PubMed]
38. Lagoumintzis, G.; Sideris, S.; Kampouris, M.; Koutosjannis, C.; Rennekampff, H.-O.; Poulas, K. Wireless Micro Current Stimulation Technology Improves Firework Burn Healing Clinical Applications of Wmcs Technology. In *Transforming Healthcare Through Innovations in Mobile and Wireless Technologies, Proceedings of the 4th International Conference on Wireless Mobile Communication and Healthcare, Athens, Greece, 3–5 November 2014*; ICST: South Portland, MM, USA, 2014.
39. Wirsing, P.G.; Konstantakaki, M.; Poulas, K.A. Martorell's ulcer successfully treated by wireless microcurrent stimulation technology. *Adv. Skin Wound Care* **2019**, *32*, 81–84. [CrossRef] [PubMed]
40. Lala, D.; Spaulding, S.J.; Burke, S.M.; Houghton, P.E. Electrical stimulation therapy for the treatment of pressure ulcers in individuals with spinal cord injury: A systematic review and meta-analysis. *Int. Wound J.* **2016**, *13*, 1214–1226. [CrossRef] [PubMed]
41. Smit, C.A.J.; de Groot, S.; Stolwijk-Swuste, J.M.; Janssen, T.W.J. Effects of electrical stimulation on risk factors for developing pressure ulcers in people with a spinal cord injury. *Am. J. Phys. Med. Rehabil.* **2016**, *95*, 535–552. [CrossRef]
42. Khouri, C.; Kotzki, S.; Roustit, M.; Blaise, S.; Gueyffier, F.; Cracowski, J.-L. Hierarchical evaluation of electrical stimulation protocols for chronic wound healing: An effect size meta-analysis. *Wound Repair Regen.* **2017**, *25*, 883–891. [CrossRef]
43. Thakral, G.; LaFontaine, J.; Najafi, B.; Talal, T.K.; Kim, P.; Lavery, L.A. Electrical stimulation to accelerate wound healing. *Diabet. Foot Ankle* **2013**, *4*, 22081. [CrossRef]
44. Ud-Din, S.; Bayat, A. Electrical stimulation and cutaneous wound healing: A review of clinical evidence. *Healthcare* **2014**, *2*, 445–467. [CrossRef] [PubMed]
45. Ashrafi, M.; Alonso-Rasgado, T.; Baguneid, M.; Bayat, A. The efficacy of electrical stimulation in lower extremity cutaneous wound healing: A systematic review. *Exp. Dermatol.* **2017**, *26*, 171–178. [CrossRef] [PubMed]
46. Kloth, L.C. Electrical stimulation technologies for wound healing. *Adv. Wound Care* **2014**, *3*, 81–90. [CrossRef] [PubMed]

47. Hunckler, J.; de Mel, A. A current affair: Electrotherapy in wound healing. *J. Multidiscip. Healthc.* **2017**, *10*, 179–194. [CrossRef] [PubMed]
48. Zhao, M.; Song, B.; Pu, J.; Wada, T.; Reid, B.; Tai, G.; Wang, F.; Guo, A.; Walczysko, P.; Gu, Y.; et al. Electrical signals control wound healing through phosphatidylinositol-3-OH kinase-γ and PTEN. *Nature* **2006**, *442*, 457–460. [CrossRef] [PubMed]

Communication

Detection of Superoxide Alterations Induced by 5-Fluorouracil on HeLa Cells with a Cell-Based Biosensor

Sophia Mavrikou [1,*], Vasileios Tsekouras [1], Maria-Argyro Karageorgou [2,3], Georgia Moschopoulou [1] and Spyridon Kintzios [1,*]

1 Faculty of Applied Biology and Biotechnology, Department of Biotechnology, Agricultural University of Athens, Iera Odos 75, 11855 Athens, Greece; tsekouras@aua.gr (V.T.); geo_mos@aua.gr (G.M.)
2 Institute of Nuclear & Radiological Sciences & Technology, Energy & Safety, National Center for Scientific Research "Demokritos", Aghia Paraskevi, 15310 Athens, Greece; kmargo@phys.uoa.gr
3 Faculty of Physics, Department of Solid State Physics, NKUA, 15784 Athens, Greece
* Correspondence: sophie_mav@aua.gr (S.M.); skin@aua.gr (S.K.); Tel.: +30-210-529-4292 (S.M.); +30-210-529-4292 (S.K.)

Received: 9 September 2019; Accepted: 13 October 2019; Published: 16 October 2019

Abstract: Background: In vitro cell culture monitoring can be used as an indicator of cellular oxidative stress for the assessment of different chemotherapy agents. Methods: A cell-based bioelectric biosensor was used to detect alterations in superoxide levels in the culture medium of HeLa cervical cancer cells after treatment with the chemotherapeutic agent 5-fluorouracil (5-FU). The cytotoxic effects of 5-fluorouracil on HeLa cells were assessed by the MTT proliferation assay, whereas oxidative damage and induction of apoptosis were measured fluorometrically by the mitochondria-targeted MitoSOX™ Red and caspase-3 activation assays, respectively. Results: The results of this study indicate that 5-FU differentially affects superoxide production and caspase-3 activation when applied in cytotoxic concentrations against HeLa cells, while superoxide accumulation is in accordance with mitochondrial superoxide levels. Our findings suggest that changes in superoxide concentration could be detected with the biosensor in a non-invasive and rapid manner, thus allowing a reliable estimation of oxidative damage due to cell apoptosis. Conclusions: These findings may be useful for facilitating future high throughput screening of different chemotherapeutic drugs with a cytotoxic principle based on free radical production.

Keywords: anticancer therapeutic strategies; apoptosis; bioelectric; biosensor; 5-fluorouracil; HeLa cell line; superoxide

1. Introduction

Gynecological cancers are life-endangering malignancies of the female reproductive system, discriminated as cervical, ovarian, uterine, vaginal, vulvar, and fallopian tube cancer after the affected anatomical organ. These diseases are estimated to cause approximately 110,000 new cases and 33,000 deaths in the United States alone on an annual basis [1].

The successful treatment of cancers requires a combination of invasive and non-invasive processes such as surgery, chemotherapy and radiation, considering efficient treatment and patient's quality of life [2]. Chemotherapy is a major constituent of the multidisciplinary cancer therapy, evolved throughout the years from a single active compound remedy to a multifactor approach, applying combined regimens towards personalized medicine. One of the most frequently used chemotherapeutic agents in cancer therapy is 5-fluorouracil (5-FU), an analogue of uracil [3,4]. The anticancer activity of 5-FU originates from the inhibition of thymidylate synthase (TS) activity during the S phase of the cell cycle and its incorporation into RNA and DNA of tumor cells [5,6]. Furthermore, 5-FU promotes cell death by generating mitochondrial ROS in the p53-dependent pathway [7–9] and by inducing apoptosis through the activation of a cascade of caspases 1, 3 and 8 [10].

Cancer cells have the ability to prevent programmed cell death by disorganizing tissue homeostasis, the balance between cell proliferation and cell death [11]. Therefore, a fundamental target of conventional chemotherapy is the activation of endocellular signaling mechanisms involved in cell death pathways, in particular, those mediating apoptosis. 5-fluorouracil is an effective pharmaceutical agent reported to initiate apoptotic processes against a number of malignancies such as colorectal [12], oral [13], breast [14], head and neck [15], gastric [16] and cervical carcinomas [17,18]. A crucial element of apoptotic cell death is caspase-3, a cysteine protease that catalyzes a number of key endocellular proteins [19]. Human caspases are a group of eleven endoproteases, caspases 1–10 and caspase 14, controlling cell regulatory networks of inflammation and cell death [20]. The executioner caspase-3 is activated from initiator caspases, the outcome of apoptotic cell signaling driven from either mitochondrial cytochrome c release or cell death receptor activation [21]. The anticancer efficacy of 5-FU, when administered alone or in combination with other agents, is associated with caspase-3 function as reported in several in vivo and in vitro studies regarding colorectal [22], gastric [23], pancreatic [24] and gynecological malignancies [25,26].

Cell culture monitoring can be used as indicator for the response to different chemotherapy options. The use of biosensors, in particular bioelectric and electrochemical sensors in the analysis of antineoplastic drugs has increased in importance over the last years [27,28]. In this context, a critical marker for monitoring cancer cells differentiation within a cell population is the superoxide anion. This molecule is mainly a by-product of the oxidative phosphorylation of the mitochondria electron transport chain. Initially, it is released to the mitochondrial matrix, where it is converted immediately to hydrogen peroxide. Mitochondrial hydrogen peroxide can then diffuse to both the cytosol and the nucleus and either interact with other free radical species, modulate signaling cascades or cause cellular damage. Together with other free radical species, superoxide has been found to mediate the development and/or survival of cancer cells and tumors, both in vivo and in vitro [29–31]. This property of superoxide has led researchers to propose the regulation of cellular redox status as a novel, critical and highly efficient cancer therapeutic strategy [32,33], utilizing superoxide dismutase along with other antioxidant systems [34,35].

Chemotherapy is associated with oxidative stress alterations that affect vital cellular processes such as cell cycle progression and drug-induced apoptosis [36–38]. The scope of this study is to investigate whether it is possible to employ a biosensor-based approach to detect in a non-invasive way the superoxide levels generated by cervical cancer cells after exposure to the anticancer agent 5-fluorouracil. For this purpose, different 5-FU concentrations were tested towards the HeLa cervical cancer line for 24 and 48 h. After the determination of 5-FU's cytotoxic activity cellular stress markers such as mitochondrial superoxide and caspase-3 levels were determined. Superoxide levels were determined in parallel in the culture medium with an advanced cell-based bioelectric biosensor, an approach which has been previously applied to monitoring superoxide levels in cultures of differentiating neuronal cells [39]. In this way, we demonstrate that it is possible to access in a high throughput, non-invasive way the in vitro efficacy of target anticancer compounds with a cytotoxic principle based on free radical production.

2. Materials and Methods

2.1. Cell Line and Culture Conditions

HeLa (ATCC® CCL-2™) and Vero cell lines were originally purchased from the American Type Culture Collection (ATCC) (Manassas, VA, USA). The cells were cultured in Dulbecco's Modified Eagle Medium (BiochromGmbh, Berlin, Germany) supplemented with 10% fetal bovine serum (Thermo Fisher Scientific, Waltham, MA, USA), 2 mM L-glutamine, 0.5 mM sodium pyruvate and 1%

penicillin/streptomycin, all procured from Biowest (Biowest, Nuaillé, France). Cells were incubated in a 5% CO_2 incubator (HF90Air jacketed CO_2 Incubator, Heal Force Bio-Meditech Holdings Limited, Shanghai, China) at 37 °C for proliferation.

2.2. MTT Cell Proliferation Assay

Cells were plated in transparent flat-bottom 96-well plates (SPL Life Sciences Co Ltd., Naechon-Myeon, Korea) at two different population densities according to two incubation time intervals (24 and 48 h): 10^3 and 8×10^3 cells per well, both supplemented with 100 µL of culture medium. The next day the medium was replaced with 200 µL of medium supplemented with 1% FBS that contained the different 5-fluorouracil (5-FU) concentrations. Cells not treated with 5-FU were considered as control (0). The cytotoxic agent 5-FU was initially diluted in dimethyl-sulfoxide. After 24 and 48 h incubation 0.5 mg/mL 3-(4,5-dimethylthiazol-2-yl)-2,5-diphenyltetrazolium bromide (MTT) (Duchefa Biochemie, Haarlem, the Netherlands) was added in each well. After 3 h in culture, the MTT-containing medium was removed and cells were solubilized with 200 µL dimethyl sulfoxide (DMSO). Cell viability was determined by measuring the absorbance at 560 nm wavelength, using a PowerWave 240 microplate photometer (Biotek, Winooski, VT, USA). The results were expressed as the percentage of absorbance values compared to control and were assessed to determine the changes in viability.

2.3. Measurement of Mitochondrial Superoxide Production

Mitochondria-targeted MitoSOX™ Red fluorogenic dye (Thermo Fisher Scientific, Rockford, IL, USA) was used to measure mitochondrial superoxide accumulation according to the manufacturer's instructions. Briefly, cells were seeded in 96-well black plates (at the same densities indicated at the MTT viability assessment) with clear bottom and were left for 24 and 48 h incubation with the bioactive compounds. Antimycin-A (50 µM) was used as a positive control. After overnight incubation, the medium was aspirated, and cells were incubated for 10 min at 37 °C in 0.2 mL of measurement buffer containing 5 µM MitoSOX™ Red. Then, the cells were washed twice with PBS. MitoSOX™ fluorescence was measured at 510 nm excitation and 580 nm emission wavelengths on an Infinite M200PRO multimode microplate reader (Tecan Group Ltd., Männedorf, Switzerland). MitoSOX™ fluorescence was adjusted based on total protein content as determined by the Bradford assay [40] measured at an absorbance of 595 nm. The results were expressed as the percentage of adjusted values compared to control.

2.4. Measurement of Caspase Activity

The activity of caspase-3 enzymes was measured in cell lysates by a colorimetric assay kit (CASP-3-C, SIGMA-ALDRICH, Saint Louis, MO, USA), that is based on the hydrolysis of the compound acetyl-Asp-Glu-Val-Asp p-nitroanilide (Ac-DEVD-pNA) (A 2559, Sigma-Aldrich, Saint Louis, MO, USA) by caspase-3. The reaction releases the p-nitroaniline moiety (p-NA) that absorbs at 405 nm. Cells were pre-incubated with 5 µM doxorubicin hydrochloride (DOX) (Sigma–Aldrich, Deisenhofen, Germany) for 3 h for caspase cascade initiation [41,42]. Samples (10 µL) were tested both with and without the caspase-3 inhibitor acetyl-Asp-Glu-Val-Asp-al (Ac-DEVD-CHO, 20 µM final concentration), in a total reaction volume of 100 µL in 96-well plates. The substrate Ac-DEVD-pNA concentration was 200 µM and the assay was performed at 37 °C for 90 min. The caspase-3 specific activity was expressed as the percentage of µmol of p-nitroaniline released per min per µg of total protein values compared to control.

2.5. HeLa Cell Total Protein Extraction

For total protein extraction, cells were lysed in assay buffer (20 mM HEPES, pH 7.4, 2 mM EDTA, 0.1% CHAPS, 5 mM DTT) containing protease inhibitors (11873580001; Roche Diagnostics; Mannheim, Germany). Protein concentration was determined by the Bradford assay [40]. Briefly, the medium was removed from cells cultured in six-well plates and cells were washed with PBS. Then, 50 μL of assay buffer was added in each well and cells were left for 20 min in −20 °C. After incubation, lysed cells were centrifuged at 15,000× *g* for 15 min at 4 °C and the supernatants were transferred to Eppendorf tubes. The cytoplasmic proteins were maintained at −80 °C until use.

2.6. Creation of Membrane-Engineered Cells and Bioensor Fabrication (Vero-SOD)

Membrane-engineered mammalian cells were created by the electroinsertion of the enzyme superoxide dismutase (SOD) into the membrane of Vero cell fibroblasts following the protocol of Moschopoulou et al. [43]. Initially, cells at a density of 3×10^6 mL^{-1} were centrifuged at 1000 rpm for 2 min and the pellet was resuspended in PBS (pH 7.4). Afterwards, cells were incubated with 1500 $U{\cdot}mL^{-1}$ CuZnSOD (EC1.15.1.1) for 20 min at 4 °C and the mixture was transferred to electroporator (Eppendorf Eporator, Eppendorf AG, Germany) cuvettes. Electroinsertion was performed by applying four pulses of an electric field at 1800 $V{\cdot}cm^{-1}$. Then, cells were centrifuged at 1000 rpm for 2 min and resuspended in cell culture medium. Finally, the sensors were fabricated by mixing 1 volume of Vero-SOD cells with 2 volumes of 4% (w/v) sodium alginate solution and was added dropwise with the use of a 22G syringe in 0.8 M $CaCl_2$. Cells were immobilized in calcium alginate, forming beads containing 75×10^3 cells per bead with an approximate diameter of 2 mm. As already reported [39,43], the membrane potential of membrane-engineered Vero cell fibroblasts is affected by the interactions of electroinserted SOD molecules and superoxide anions, producing measurable changes in the membrane potential.

2.7. Biosensor Setup for Recording Superoxide Concentration and Data Processing

For recording the signal and processing of data, the PMD-1608FSA/D card (Measurement Computing, Norton, MA, USA) recording device and the software InstaCal (Measurement Computing) were used, respectively. A two-electrode system (working and reference) was connected to the device. These silver electrodes were electrochemically coated with an AgCl layer. A cell-bearing bead was attached to the working electrode while a cell-free bead was connected to the reference electrode. For each assay, both beads (sensor system, Figure 1) were immersed into the well containing adherent cells [39] and the response of each biosensor potential was achieved within 100 s after its sinking into the culture medium. The biosensor was calibrated with known superoxide concentration produced by the oxidation of xanthine by the xanthine oxidase. The range of xanthine concentration was from 1 pM to 10 nM and the xanthine oxidase was 100 mU/mL [43]. Each response was expressed as the average of the cellular membrane potential of each assay, which has been calibrated to correspond to relative changes in superoxide concentration.

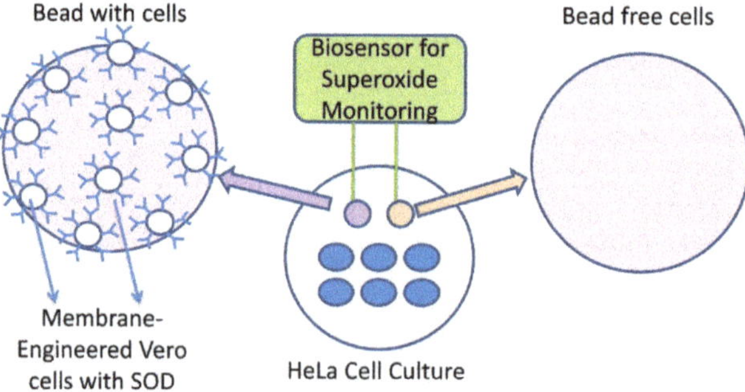

Figure 1. Cell-based biosensor system configuration.

2.8. Statistical Analysis

Each experiment was repeated independently three times for each treatment with n = 5. Significance testing in comparisons was based on Student's *t*-tests for pairs. *p*-values < 0.05 were considered to be statistically significant.

3. Results

3.1. Assessment of the Effects of the Chemotherapeutic Agent 5-Fluorouracil (5-FU) on HeLa Cell Viability

In order to study the 5-fluorouracil-induced cytotoxicity, HeLa cells were initially exposed to various concentrations of the cytostatic agent for 24 and 48 h. Doses up to 200 μM were used causing 30% and 49% inhibition of cell growth at 24 (Figure 2a) and 48 h (Figure 2b), respectively (at the highest dose of 200 μM).

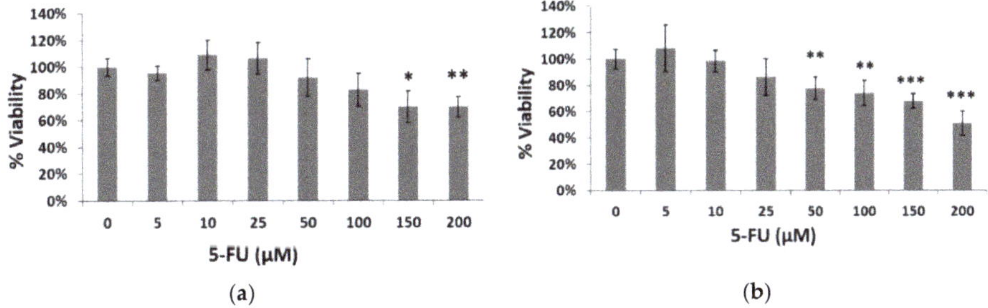

Figure 2. Percentage of HeLa cell viable cells after treatment with 5-fluorouracil (5-FU) at concentrations 5, 10, 25, 50, 100, 150 and 200 μM for (**a**) 24 h and (**b**) 48 h. Average results from replicate experiments ± SD ($n = 3$). * $p < 0.05$, ** $p < 0.01$, *** $p < 0.001$, significantly different from the control.

3.2. Increased Mitochondrial Superoxide Production in HeLa Cells is Observed after 24 h Treatment with 5-FU

For the determination of mitochondrial superoxide production, cells were loaded with the fluorescent probe MitoSOX™ Red after exposure with a standard lethal 5-FU concentration (150 μM) for 24 h and 48 h (Figure 3). Our results indicated an increase in the mitochondrial superoxide levels in comparison with the control after 24 h cell exposure with 5-FU whereas a 48 h incubation led to a decrease in superoxide accumulation, possibly associated with cell loss due to 5-FU toxicity.

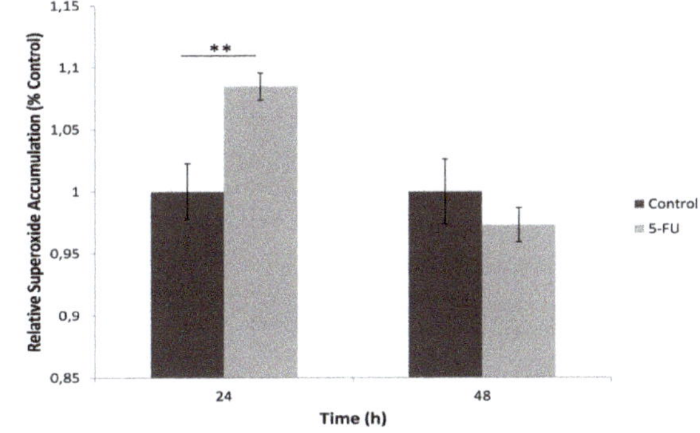

Figure 3. Mitochondrial superoxide levels in HeLa cells after treatment with 5-FU (150 μM) for 24 and 48 h, assessed as MitoSOX™ Red fluorescence intensity normalized to total protein content and control (no treatment with 5-FU). Average results from replicate experiments ± SD ($n = 3$). ** $p < 0.01$, significantly different from the control.

3.3. Caspase-3 Activation after 48 h Treatment with 5-FU

In order to determine the effect of 5-FU on the activation of caspase-3, HeLa cells were once again treated with 5-FU for 24 and 48 h and then the caspase specific activity was measured (Figure 4). HeLa cells treated for 24 h presented no significant caspase activity. On the contrary, caspase activation was remarkably increased at 48 h after treatment with 5-FU. Similar to mitochondrial superoxide levels, this observation was possibly associated with increased cell death after the 48-h treatment with 5-FU.

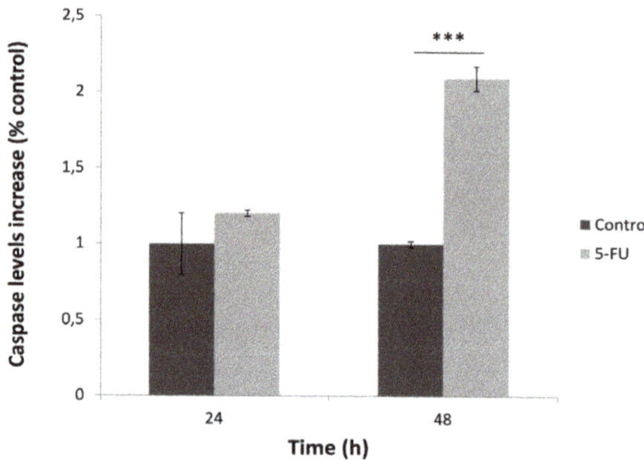

Figure 4. Caspase-3 specific activity in HeLa cells after treatment with 5-FU (150 μM) for 24 h and 48 h, normalized to control (no treatment with 5-FU). Average results from replicate experiments ± SD ($n = 3$). *** $p < 0.001$, significantly different from the control.

3.4. Superoxide Accumulation in Cell Culture Determined by the Bioelectric Cell-Based Biosensor

A significantly increased superoxide concentration determined by the bioelectric biosensor-based assay was observed following the 24 h exposure to 5-FU compared to the control (Figure 5). On the other hand, we observed a significant reduction of superoxide levels after the 48 h treatment with the anticancer agent. The results of superoxide determination in the culture medium with the biosensor assay were very highly correlated with the respective results of the MitoSOX™ Red mitochondrial superoxide assay, though positively for the 24 h treatment ($r^2 = 0.99$) and negatively for the 48 h treatment ($r^2 = -0.99$).

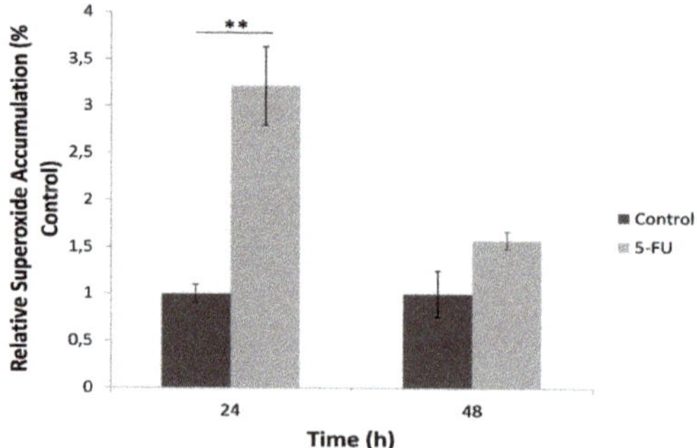

Figure 5. Relative superoxide accumulation in the culture medium during the 24 and 48 h incubation of HeLa cells with 5-FU, as determined with the cell-based superoxide biosensor, normalized to control (no treatment with 5-FU). Average results from replicate experiments ± SD ($n = 3$). ** $p < 0.01$, significantly different from the control.

4. Discussion

In this study, we demonstrate the successful implementation of a cell-based biosensor setup for the direct and rapid detection of superoxide released from cervical cancer cell culture treated with 5-FU, one of the most widely applied chemotherapeutic compounds in clinical therapy. Superoxide is considered a particularly significant ROS with an extremely short half-life; thus real-time level monitoring is an important yet complex issue. The in vitro determination of reactive oxygen species superoxide is generally assessed by spectrophotometric, analytical, electron spin resonance and electrochemical biosensor approaches [44]. Biosensors could provide reliable methods for the detection of relative superoxide levels in cell cultures without the need for extensive chemical and physical cell culture treatments.

The determination of dynamic radical changes in cellular microenvironment with non-invasive approaches is an essential mean for assessing the effectiveness of anticancer agents on cancer treatment [45–48]. As mentioned above, 5-FU's chemotherapeutic properties are attributed to thymidylate synthase (TS) activity inhibition, to mitochondrial ROS generation and the caspases cascade activation [49]. Our findings demonstrate that 5-FU differentially affects superoxide production and caspase-3 activation when applied in cytotoxic concentrations against HeLa cells. MTT assay was used to evaluate the antiproliferative effect of 5-FU at 24 h and 48 h treatments. We spectrophotometrically determined elevated mitochondrial superoxide production after the 24-h treatment but no changes in caspase-3 activity. On the contrary, mitochondrial superoxide release was decreased to control levels whereas caspase-3 activity was significantly upregulated. These trends are in accordance with the results obtained from the superoxide biosensor as 5-FU significantly upraised superoxide production after the 24 h treatment, followed by a downregulation to control levels after 48-h incubation, possibly also associated with increased number of cells undergoing apoptosis.

The correlation between the response of the superoxide bioelectric biosensor and the superoxide concentration is indirectly linked to changes caused in the cell membrane potential of cells which have been membrane-engineered with SOD moieties, i.e., SOD units which have been electroinserted in the cell membrane. This technology is known as molecular identification through membrane engineering. This is a generic methodology of artificially inserting tens of thousands of receptor molecules on the cell surface, thus rendering the cell a selective responder against analytes binding to the inserted receptors. Receptor molecules can vary from antibodies to enzymes to polysaccharides [50–54]. It has been previously proven [55] that this principle is associated, in a unique way, with certain changes in the cellular electric properties (in particular, cell membrane hyperpolarization) as a consequence of the interaction of the analytes under determination with the electroinserted molecules and therefore changes in the cellular structure.

It must be emphasized that cell membrane hyperpolarization is the dominant change in the engineered cell membrane electric properties following a mechanical distortion of the membrane (according to the novel assay principle described in the method) as also expected by the concurrent change in the actin cytoskeleton structure, in particular the actin cytoskeleton network adjunct to the sites of the interaction, including the circumferential actin belt and changes in the propagation of electric signals along actin filaments [56]. On the contrary, the main change in the cell membrane electric properties of normal (i.e., non-membrane engineered) cells is depolarization, as frequently described in prior approaches [57–59].

It has been previously proven that superoxide dismutation triggered changes to the membrane potential of fibroblast cells membrane-engineered with SOD [43]. In order to measure the aforementioned changes in the engineered cell membrane potential, the superoxide biosensor used in the present study was designed according to the principle of the bioelectric recognition assay (BERA) [57]. This methodology has been often applied to detect cellular interactions with bioactive compounds via the determination of electrical conductivity generated adjacent to a cell cluster as a means of indirect measurement of the relative changes in the cell membrane potential. This means that the AgCl electrode system applied is able to measure conductivity alterations of the extracellular microenvironment reflecting the bioelectric profiling of cell responses to various treatments [60]. It has been previously demonstrated that by using this superoxide biosensor system it is possible to rapidly measure superoxide concentrations as low as 1 pM [43] and also to correlate the biosensor response with currently available conventional methods for superoxide determination [43,50,61].

Mitochondrial superoxide is produced when electrons released from the electron transfer system in the inner membrane of mitochondria are captured by molecular oxygen and become superoxide [62]. It has been reported that extracellular superoxide is closely linked to mitochondrial superoxide production [63]. Our results support this statement as high mitochondrial superoxide levels, measured with mitochondria-targeted MitoSOX™ Red fluorogenic dye are accompanied by high extracellular accumulation of the anion. However, superoxide concentration data recorded with the cell-based biosensor should be considered with caution compared with MitoSOX™ results, since the biosensor assayed superoxide eluted in the culture medium, not its actual intracellular concentration. That said, it is worth elaborating on the observation that, although the results obtained with the bioelectric biosensor for the 24 h treatment were very highly and positively correlated with the MitoSOX™ assay, a negative yet high correlation was established for the 48-h treatment. This could be possibly explained by the reduced number of live HeLa cells after their prolonged exposure to 5-FU. As already hypothesized in Section 3, this would result in a lower level of superoxide accumulation in the mitochondria (since fewer cells will be metabolically active) and at the same time in a higher level of superoxide released in the culture media (compared to control) due to cell membrane leakage from dead cells. In this respect, the reliability of the biosensor-based assay in accessing the ROS-mediated toxicity of 5-FU is considerably higher than the MitoSOX™ assay, since it allows for a more realistic and accurate estimation of the level of oxidative stress related to and/or resulting from the anticancer drug effect.

5. Conclusions

Intracellular ROS are increasingly recognized as critical determinants of cellular signaling and can be used as markers of 5-FU-mediated anti-tumor effects. Biosensors are widely used for the assessment of physiological and pathological applications since they are non-invasive methods for the detection of crucial biomarkers of cancer cell cytotoxicity without the need for extensive sample pretreatment. In this study, we present a cell-based biosensor tool suitable for cell culture superoxide monitoring without the need for cell transfer and staining, which could affect cellular responses due to extrinsic chemical and mechanical stress. This biosensor is applied for the estimation of total superoxide levels released from the cancer cells into the culture medium after treatment with the chemotherapeutic agent. Our results are in agreement with the findings of the biochemical assays for the assessment of mitochondrial superoxide and caspase-3 activities. Further validation is required of the feasibility of our biosensor-based approach for monitoring alterations in the in vitro redox status of cancer cells treated with anticancer agents, e.g., by testing more and different cancer cell lines crossed with chemotherapeutical drug combinations. In a longer perspective, the availability of a reliable, high throughput and rapid system for superoxide determination in cancer cells will allow for the accurate assessment of chemoresistance in cervical and other cancer cells, at least as far as its association with redox balance is concerned [38,64–66].

Author Contributions: Methodology, S.M., V.T., M.-A.K., G.M.; investigation, S.M., V.T., M.-A.K., G.M.; data curation, S.M., V.T.; writing—original draft preparation, S.M., V.T., S.K., G.M.; writing—review and editing, S.K.; supervision, S.K.; project administration, G.M.

Funding: This work is supported by the Operational Program "Human Resources Development, Education and Lifelong Learning", Priority Axes 6, 8, 9 co-financed by European Social Fund entitled "Support of Researchers with Emphasis on Senior Researchers" (NSRF 2014-2020).

Conflicts of Interest: The authors declare no conflict of interest.

References

1. Siegel, R.L.; Miller, K.D.; Jemal, A. Cancer statistics. *CA Cancer J. Clin.* **2019**, *69*, 7–34. [CrossRef] [PubMed]

2. Melville, A.; Eastwood, A.; Kleijnen, J.; Kitchener, H.; Martin-Hirsch, P.; Nelson, L. Management of gynaecological cancers. *Qual. Health Care* **1999**, *8*, 270–279. [CrossRef] [PubMed]
3. Grem, J.L. Mechanisms of Action and Modulation of Fluorouracil. *Semin. Radiat. Oncol.* **1997**, *7*, 249–259. [CrossRef]
4. Zhang, Y.; Talmon, G.; Wang, J. MicroRNA-587 antagonizes 5-FU-induced apoptosis and confers drug resistance by regulating PPP2R1B expression in colorectal cancer. *Cell Death Dis.* **2015**, *6*, e1845. [CrossRef]
5. Noordhuis, P.; Holwerda, U.; van der Wilt, C.L.; Groeningen, C.; Smid, K.; Meijer, S.; Pinedo, H.; Peters, G. 5-Fluorouracil incorporation into RNA and DNA in relation to thymidylate synthase inhibition of human colorectal cancers. *Ann. Oncol.* **2004**, *15*, 1025–1032. [CrossRef]
6. Walko, C.M.; Lindley, C. Capecitabine: A review. *Clin. Ther.* **2005**, *27*, 23–44. [CrossRef]
7. Hwang, P.M.; Bunz, F.; Yu, J.; Rago, C.; Chan, T.A.; Murphy, M.P.; Kelso, G.F.; Smith, R.A.; Kinzler, K.W.; Vogelstein, B. Ferredoxin reductase affects p53-dependent, 5-fluorouracil-induced apoptosis in colorectal cancer cells. *Nat. Med.* **2001**, *7*, 1111–1117. [CrossRef]
8. Fan, C.; Chen, J.; Wang, Y.; Wong, Y.S.; Zhang, Y.; Zheng, W.; Cao, W.; Chen, T. Selenocystine potentiates cancer cell apoptosis induced by 5-fluorouracil by triggering reactive oxygen species-mediated DNA damage and inactivation of the ERK pathway. *Free Radic. Biol. Med.* **2013**, *65*, 305–316. [CrossRef]
9. Liu, M.P.; Liao, M.; Dai, C.; Chen, J.F.; Yang, C.J.; Liu, M.; Chen, Z.G.; Yao, M.C. Sanguisorba officinalis L. synergistically enhanced 5-fluorouracil cytotoxicity in colorectal cancer cells by promoting a reactive oxygen species-mediated, mitochondria-caspase-dependent apoptotic pathway. *Sci. Rep.* **2016**, *27*, 34245. [CrossRef]
10. Ohtani, T.; Hatori, M.; Ito, H.; Takizawa, K.; Kamijo, R.; Nagumo, M. Involvement of caspases in 5-FU induced apoptosis in an oral cancer cell line. *Anticancer Res.* **2000**, *20*, 3117–3121.
11. Ricci, M.S.; Zong, W.X. Chemotherapeutic approaches for targeting cell death pathways. *Oncologist* **2006**, *11*, 342–357. [CrossRef] [PubMed]
12. Mhaidat, N.M.; Bouklihacene, M.; Thorne, R.F. 5-Fluorouracil-induced apoptosis in colorectal cancer cells is caspase-9-dependent and mediated by activation of protein kinase C-δ. *Oncol. Lett.* **2014**, *8*, 699–704. [CrossRef] [PubMed]
13. Tong, D.; Poot, M.; Hu, D.; Oda, D. 5-Fluorouracil-induced apoptosis in cultured oral cancer cells. *Oral Oncol.* **2000**, *36*, 236–241. [CrossRef]
14. Ponce-Cusi, R.; Calaf, G. Apoptotic activity of 5-fluorouracil in breast cancer cells transformed by low doses of ionizing α-particle radiation. *Int. J. Oncol.* **2016**, *48*, 774–782. [CrossRef] [PubMed]
15. Shigeishi, H.; Biddle, A.; Gammon, L.; Rodini, C.O.; Yamasaki, M.; Seino, S.; Sugiyama, M.; Takechi, M.; Mackenzie, I.C. Elevation in 5-FU-induced apoptosis in head and neck cancer stem cells by a combination of CDHP and GSK3β inhibitors. *J. Oral Pathol. Med.* **2015**, *44*, 201–207. [CrossRef]
16. Zhao, P.; Zhang, J.; Pang, X.; Zhao, L.; Li, Q.; Cao, B. Effect of apatinib combined with 5-fluorouracil (5-FU) on proliferation, apoptosis and invasiveness of gastric cancer cells. *Ann. Oncol.* **2017**, *28*. [CrossRef]
17. Koraneekit, A.; Limpaiboon, T.; Sangka, A.; Boonsiri, P.; Daduang, S.; Daduang, J. Synergistic effects of cisplatin-caffeic acid induces apoptosis in human cervical cancer cells via the mitochondrial pathways. *Oncol. Lett.* **2018**, *15*, 7397–7402. [CrossRef]
18. Hemaiswarya, S.; Doble, M. Combination of phenylpropanoids with 5-fluorouracil as anti-cancer agents against human cervical cancer (HeLa) cell line. *Phytomedicine* **2012**, *20*, 151–158. [CrossRef]
19. Porter, A.G.; Jänicke, R.U. Emerging roles of caspase-3 in apoptosis. *Cell Death Differ.* **1999**, *6*, 99–104. [CrossRef]
20. Li, J.; Yuan, J. Caspases in apoptosis and beyond. *Oncogene* **2008**, *27*, 6194–6206. [CrossRef]
21. Parrish, A.B.; Freel, C.D.; Kornbluth, S. Cellular mechanisms controlling caspase activation and function. *Cold Spring Harb. Perspect. Biol.* **2013**, *5*, a008672. [CrossRef] [PubMed]
22. Shi, T.; Gao, M.; He, M.; Yue, F.; Zhao, Y.; Sun, M.; He, K.; Chen, L. 5-FU preferably induces apoptosis in BRAF V600E colorectal cancer cells via downregulation of Bcl-xL. *Mol. Cell. Biochem.* **2019**, *461*, 151–158. [CrossRef] [PubMed]
23. Fu, Z.; Zhou, Q.; Zhu, S.; Liu, W. Anti-tumor mechanism of IL-21 used alone and in combination with 5-fluorouracil in vitro on human gastric cancer cells. *J. Biol. Regul. Homeost. Agents* **2018**, *32*, 619–625. [PubMed]

24. Ghadban, T.; Dibbern, J.L.; Reeh, M.; Miro, J.T.; Tsui, T.Y.; Wellner, U.; Izbicki, J.R.; Güngör, C.; Vashist, Y.K. HSP90 is a promising target in gemcitabine and 5-fluorouracil resistant pancreatic cancer. *Apoptosis* **2017**, *22*, 369–380. [CrossRef] [PubMed]
25. Chen, X.X.; Leung, G.P.H.; Zhang, Z.J.; Xiao, J.B.; Lao, L.X.; Feng, F.; Zhang, K.Y.B. Proanthocyanidins from Uncaria rhynchophylla induced apoptosis in MDA-MB-231 breast cancer cells while enhancing cytotoxic effects of 5-fluorouracil. *Food Chem. Toxicol.* **2017**, *107*, 248–260. [CrossRef]
26. Deveci, H.A.; Nazıroğlu, M.; Nur, G. 5-Fluorouracil-induced mitochondrial oxidative cytotoxicity and apoptosis are increased in MCF-7 human breast cancer cells by TRPV1 channel activation but not Hypericum perforatum treatment. *Mol. Cell. Biochem.* **2018**, *439*, 189–198. [CrossRef]
27. Lima, H.R.S.; da Silva, J.S.; de Oliveira Farias, E.A.; Teixeira, P.R.S.; Eiras, C.; Nunes, L.C.C. Electrochemical sensors and biosensors for the analysis of antineoplastic drugs. *Biosens. Bioelectron.* **2018**, *108*, 27–37. [CrossRef]
28. Meneghello, A.; Tartaggia, S.; Alvau, M.D.; Polo, F.; Toffoli, G. Biosensing technologies for therapeutic drug monitoring. *Curr. Med. Chem.* **2018**, *25*, 4354–4377. [CrossRef]
29. Shibayama-Imazu, T.; Sonoda, I.; Sakairi, S.; Aiuchi, T.; Ann, W.W.; Nakajo, S.; Itabe, H.; Nakaya, K. Production of superoxide and dissipation of mitochondrial transmembrane potential by vitamin K2 trigger apoptosis in human ovarian cancer TYK-nu cells. *Apoptosis* **2006**, *11*, 1535–1543. [CrossRef]
30. Dhar, S.K.; St Clair, D.K. Manganese superoxide dismutase regulation and cancer. *Free Radic. Biol. Med.* **2012**, *52*, 2209–2222. [CrossRef]
31. Coso, S.; Harrison, I.; Harrison, C.B.; Vinh, A.; Sobey, C.G.; Drummond, G.R.; Williams, E.D.; Selemidis, S. NADPH oxidases as regulators of tumor angiogenesis: Current and emerging concepts. *Antioxid. Redox Signal.* **2012**, *16*, 1229–1247. [CrossRef] [PubMed]
32. Baffy, G.; Derdak, Z.; Robson, S.C. Mitochondrial recoupling: A novel therapeutic strategy for cancer? *Br. J. Cancer* **2011**, *105*, 469–474. [CrossRef]
33. Montero, A.J.; Jassem, J. Cellular redox pathways as a therapeutic target in the treatment of cancer. *Drugs* **2011**, *71*, 1385–1396. [CrossRef] [PubMed]
34. Brown, D.P.; Chin-Sinex, H.; Nie, B.; Mendonca, M.S.; Wang, M. Targeting superoxide dismutase 1 to overcome cisplatin resistance in human ovarian cancer. *Cancer Chemother. Pharmacol.* **2009**, *63*, 723–730. [CrossRef] [PubMed]
35. Radenkovic, S.; Milosevic, Z.; Konjevic, G.; Karadzic, K.; Rovcanin, B.; Buta, M.; Gopcevic, K.; Jurisic, V. Lactate dehydrogenase, catalase, and superoxide dismutase in tumor tissue of breast cancer patients in respect to mammographic findings. *Cell Biochem. Biophys.* **2013**, *66*, 287–295. [CrossRef]
36. Shacter, E.; Williams, J.A.; Hinson, R.M.; Senturker, S.; Lee, Y.-J. Oxidative stress interferes with cancer chemotherapy: Inhibition of lymphoma cell apoptosis and phagocytosis. *Blood* **2000**, *96*, 307–313. [CrossRef]
37. Hu, Y.; Rosen, D.G.; Zhou, Y.; Feng, L.; Yang, G.; Liu, J.; Huang, P. Mitochondrial manganese-superoxide dismutase expression in ovarian cancer: Role in cell proliferation and response to oxidative stress. *J. Biol. Chem.* **2005**, *280*, 39485–39492. [CrossRef]
38. Tamaki, R.; Kanai-Mori, A.; Morishige, Y.; Koike, A.; Yanagihara, K.; Amano, F. Effects of 5-fluorouracil, adriamycin and irinotecan on HSC-39, a human scirrhous gastric cancer cell line. *Oncol. Rep.* **2017**, *37*, 2366–2374. [CrossRef]
39. Moschopoulou, G.; Kintzios, S. Non-invasive Superoxide Monitoring of In Vitro Neuronal Differentiation Using a Cell-Based Biosensor. *J. Sens.* **2015**, *2015*, 768352. [CrossRef]
40. Bradford, M.M. A rapid and sensitive method for the quantitation of microgram quantities of protein utilizing the principle of protein-dye binding. *Anal. Biochem.* **1976**, *72*, 248–254. [CrossRef]
41. Ueno, M.; Kakinuma, Y.; Yuhki, K.; Murakoshi, N.; Iemitsu, M.; Miyauchi, T.; Yamaguchi, I. Doxorubicin induces apoptosis by activation of caspase-3 in cultured cardiomyocytes in vitro and rat cardiac ventricles in vivo. *J. Pharmacol. Sci.* **2006**, *101*, 151–158. [CrossRef] [PubMed]
42. Wang, S.; Konorev, E.A.; Kotamraju, S.; Joseph, J.; Kalivendi, S.; Kalyanaraman, B. Doxorubicin Induces Apoptosis in Normal and Tumor Cells via Distinctly Different Mechanisms. *J. Biol. Chem.* **2004**, *279*, 25535–25543. [CrossRef] [PubMed]
43. Moschopoulou, G.; Valero, T.; Kintzios, S. Superoxide determination using membrane-engineered cells: An example of a novel concept for the construction of cell sensors with customized target recognition properties. *Sens. Actuators B Chem.* **2012**, *175*, 78–84. [CrossRef]

44. Zhang, Y.; Dai, M.; Yuan, Z. Methods for the detection of reactive oxygen species. *Anal. Methods* **2018**, *10*, 4625–4638. [CrossRef]
45. Senthil, K.; Aranganathan, S.; Nalini, N. Evidence of oxidative stress in the circulation of ovarian cancer patients. *Clin. Chim. Acta* **2004**, *339*, 27–32. [CrossRef] [PubMed]
46. Patel, S.; Nanda, R.; Sahoo, S.; Mohapatra, E. Biosensors in Health Care: The Milestones Achieved in Their Development towards Lab-on-Chip-Analysis. *Biochem. Res. Int.* **2016**, *2016*, 3130469. [CrossRef]
47. Lewandowski, M.; Gwozdzinski, K. Nitroxides as antioxidants and anticancer drugs. *Int. J. Mol. Sci.* **2017**, *18*, 2490. [CrossRef]
48. Nunes, S.C.; Serpa, J. Glutathione in ovarian cancer: A double-edged sword. *Int. J. Mol. Sci.* **2018**, *19*, 1882. [CrossRef]
49. Akpinar, B.; Bracht, E.V.; Reijnders, D.; Safarikova, B.; Jelinkova, I.; Grandien, A.; Vaculova, A.H.; Zhivotovsky, B.; Olsson, M. 5-Fluorouracil-induced RNA stress engages a TRAIL-DISC-dependent apoptosis axis facilitated by p53. *Oncotarget* **2015**, *6*, 43679–43697. [CrossRef]
50. Moschopoulou, G.; Kintzios, S. Application of "membrane-engineering" to bioelectric recognition cell sensors for the detection of picomole concentrations of superoxide radical: A novel biosensor principle. *Anal. Chim. Acta* **2006**, *573–574*, 90–96. [CrossRef]
51. Moschopoulou, G.; Vitsa, K.; Bem, F.; Vassilakos, N.; Perdikaris, A.; Blouhos, P.; Yialouris, C.; Frossiniotis, D.; Anthopoulos, I.; Maggana, O.; et al. Engineering of the membrane of fibroblast cells with virus-specific antibodies: A novel biosensor tool for virus detection. *Biosens. Bioelectron.* **2008**, *24*, 1033–1036. [CrossRef] [PubMed]
52. Perdikaris, A.; Alexandropoulos, N.; Kintzios, S. Development of a Novel, Ultra-rapid Biosensor for the Qualitative Detection of Hepatitis B Virus-associated Antigens and Anti-HBV, Based on "Membrane-engineered" Fibroblast Cells with Virus-Specific Antibodies and Antigens. *Sensors* **2009**, *9*, 2176–2186. [CrossRef] [PubMed]
53. Perdikaris, A.; Vassilakos, N.; Yiakoumettis, I.; Kektsidou, O.; Kintzios, S. Development of a portable, high throughput biosensor system for rapid plant virus detection. *J. Virol. Methods* **2011**, *177*, 94–99. [CrossRef] [PubMed]
54. Gramberg, B.; Kintzios, S.; Schmidt, U.; Mewis, I.; Ulrichs, C. A basic approach towards the development of bioelectric bacterial biosensors for the detection of plant viruses. *J. Phytopathol.* **2012**, *160*, 106–111. [CrossRef]
55. Kokla, A.; Blouchos, P.; Livaniou, E.; Zikos, C.; Kakabakos, S.E.; Petrou, P.S.; Kintzios, S. Visualization of the membrane-engineering concept: Evidence for the specific orientation of electroinserted antibodies and selective binding of target analytes. *J. Mol. Recognit.* **2013**, *26*, 627–632. [CrossRef]
56. Nin, V.; Hernandez, J.A.; Chiffle, S. Hyperpolarization of the plasma membrane potential provokes reorganization of the actin cytoskeleton and increases the stability of adherens junctions in bovine corneal endothelial cells in culture. *Cell Motil. Cytoskelet.* **2009**, *66*, 1087–1099. [CrossRef]
57. Kintzios, S.; Bem, F.; Mangana, O.; Nomikou, K.; Markoulatos, P.; Alexandropoulos, N.; Fasseas, C.; Arakelyan, V.; Petrou, A.-L.; Soukouli, K.; et al. Study on the mechanism of Bioelectric Recognition Assay: Evidence for immobilized cell membrane interactions with viral fragments. *Biosens. Bioelectron.* **2004**, *20*, 907–916. [CrossRef]
58. Mavrikou, S.; Flampouri, E.; Moschopoulou, G.; Mangana, O.; Michaelides, A.; Kintzios, S. Assessment of organophosphate and carbamate pesticide residues in cigarette tobacco with a novel cell biosensor. *Sensors* **2008**, *8*, 2818–2832. [CrossRef]
59. Flampouri, E.; Mavrikou, S.; Kintzios, S.; Miliaids, G.; Aplada-Sarli, P. Development and Validation of a Cellular Biosensor Detecting Pesticide Residues in Tomatoes. *Talanta* **2010**, *80*, 1799–1804. [CrossRef]
60. Crowe, S.M.; Kintzios, S.; Kaltsas, G.; Palmer, C.S. A Bioelectronic System to Measure the Glycolytic Metabolism of Activated CD4+ T Cells. *Biosensors* **2019**, *9*, 10. [CrossRef]
61. Moschopoulou, G.; Papanastasiou, I.; Makri, O.; Lambrou, N.; Economou, G.; Soukouli, K.; Kintzios, S. Cellular redox-status is associated with regulation of frond division in *Spirodela polyrrhiza*. *Plant Cell Rep.* **2007**, *26*, 2063–2069. [CrossRef] [PubMed]
62. Suski, J.M.; Lebiedzinska, M.; Bonora, M.; Pinton, P.; Duszynski, J.; Wieckowski, M.R. Relation between mitochondrial membrane potential and ROS formation. *Methods Mol. Biol.* **2012**, *810*, 183–250. [CrossRef] [PubMed]

63. Li, Y.; Zhu, H.; Kuppusamy, P.; Zweier, J.L.; Trush, M.A. Mitochondrial Electron Transport Chain-Derived Superoxide Exits Macrophages: Implications for Mononuclear Cell-Mediated Pathophysiological Processes. *React. Oxyg. Species* **2016**, *1*, 81–98. [CrossRef] [PubMed]
64. Chen, J.; Solomides, C.; Parekh, H.; Simpkins, F.; Simpkins, H. Cisplatin resistance in human cervical, ovarian and lung cancer cells. *Cancer Chemother. Pharmacol.* **2015**, *75*, 1217–1227. [CrossRef]
65. Liu, Y.; Li, Q.; Zhou, L.; Xie, N.; Nice, E.C.; Zhang, H.; Huang, C.; Lei, Y. Cancer drug resistance: Redox resetting renders a way. *Oncotarget* **2016**, *7*, 42740–42761. [CrossRef]
66. Luo, M.; Wicha, M.S. Targeting cancer stem cell redox metabolism to enhance therapy responses. *Semin. Radiat. Oncol.* **2019**, *29*, 42–54. [CrossRef]

Article

Bioelectrical Analysis of Various Cancer Cell Types Immobilized in 3D Matrix and Cultured in 3D-Printed Well

Georgia Paivana [1], Sophie Mavrikou [1,*], Grigoris Kaltsas [2] and Spyridon Kintzios [1]

1 Laboratory of Cell Technology, Department of Biotechnology, Agricultural University of Athens, 118 55 Athens, Greece; georpaiv@gmail.com (G.P.); skin@aua.gr (S.K.)

2 microSENSES Laboratory, Department of Electrical and Electronic Engineering, University of West Attica, 122 44 Athens, Greece; G.Kaltsas@uniwa.gr

* Correspondence: sophie_mav@aua.gr; Tel.: +30-210-529-4292

Received: 3 October 2019; Accepted: 11 November 2019; Published: 14 November 2019

Abstract: Cancer cell lines are important tools for anticancer drug research and assessment. Impedance measurements can provide valuable information about cell viability in real time. This work presents the proof-of-concept development of a bioelectrical, impedance-based analysis technique applied to four adherent mammalian cancer cells lines immobilized in a three-dimensional (3D) calcium alginate hydrogel matrix, thus mimicking in vivo tissue conditions. Cells were treated with cytostatic agent5-fluoruracil (5-FU). The cell lines used in this study were SK-N-SH, HEK293, HeLa, and MCF-7. For each cell culture, three cell population densities were chosen (50,000, 100,000, and 200,000 cells/100 μL). The aim of this study was the extraction of mean impedance values at various frequencies for the assessment of the different behavior of various cancer cells when 5-FU was applied. For comparison purposes, impedance measurements were implemented on untreated immobilized cell lines. The results demonstrated not only the dependence of each cell line impedance value on the frequency, but also the relation of the impedance level to the cell population density for every individual cell line. By establishing a cell line-specific bioelectrical behavior, it is possible to obtain a unique fingerprint for each cancer cell line reaction to a selected anticancer agent.

Keywords: cell immobilization; 3D-printed well; bioelectric profiling; impedance analysis; real-time measurements

---

1. Introduction

Cancer is the main cause of death in many countries, as it appears in different types, most commonly affecting women (e.g., cervical, breast, and lung adenocarcinoma cancers) [1,2]. In medical treatment, doctors take steps to anticipate the development of the disease (primary prophylaxis) or to minimize its further development (secondary prophylaxis) [3]. Considering secondary prophylaxis measures, sophisticated processes are required to detect possible cellular disorders at the very earliest stages of the disease's incubation period, taking into account the dependence of the timeliness with which the disease is detected [4]. Due to the fact that many cancer diagnostic methods combined with radiological, surgical biopsy, and pathological assessments of tissue samples based on immunohistochemical and morphological characteristics [5] are time-consuming, invasive, and complicated, and require rigorous laboratory conditions, new cancer detection methods are being developed which are minimally invasive, more reliable, cheaper, and easier to use [6].

Chemotherapy, newer immunotherapy, and targeted therapy constitute the various treatment strategies that have been proposed and modified in order to increase effectiveness and precision [7,8]. Although many approaches for cancer treatments have been developed successfully, sooner or later, resistance among the subgroups of cancer cells will emerge as a hurdle to the efficacy of most current therapeutic approaches [9]. One of the most commonly used drugs for cancer treatment is 5-fluorouracil (5-FU). This compound is used in the treatment of many types of cancer, including breast cancer, colon cancer, skin cancer, etc., as it intercalates in nucleoside metabolism, leading to cytotoxicity and cell death [10].

Electrical impedance spectroscopy (EIS) is a technique that measures the electrical impedance of living cells in order to identify various cell types. This technique can be used for the successful separation of pathological cells from normal ones, taking into account the electrophysiological properties of cells based on the frequency range [11,12]. Cell electrical impedance can identify the physical, mechanical, and biochemical functions of living biological cells. EIS focuses on the analysis and discrimination of cancer cells, utilizing the fact that impedance measurements represent an effective approach for cell characterization based on the electrical responses over a particular frequency domain. As the impedance value of several tissue parameters (e.g., morphology, growth, and differentiation) vary with the frequency of the applied signal, an impedance analysis conducted over a wide frequency range provides more information about the tissue interiors, which helps us to better understand the biological tissues physiology, anatomy, and pathology [13]. For example, researchers have used the impedance technique to measure three dimensional cell cultures for four breast cell lines [14–16].

Recently, bioimpedance has been able to provide in-depth biological measurement analyses from the cell-level to DNA [17,18]. The evaluation of parameters such as cell adhesion, differentiation, spreading, morphology, growth, motility, and death for any adherent cell type is possible by monitoring the impedance changes at the contact point between the cells and electrodes [19]. In addition, bioimpedance research can nominate the pathological status of a single cell, and also be used to determine the occurrence of bacterial infections, toxicity, and changes of environmental parameters, and in the direct or indirect detection of compounds and other factors [20]. Critical changes in cellular behavior, such as the integrity of the extracellular membrane, morphology, as well as alterations in intracellular structure, significantly influence the corresponding impedance level which can be detected quickly and cost-effectively using electrodes [13,19]. Thus, impedance measurements can also be used in studying cell viability, which provides an alternative to slow and invasive traditional cytotoxicity assays [21].

Biomaterial research for drug development, cell culture, and tissue regeneration applications aims to mimic the natural extracellular matrix (ECM) in order to bridge the gap between in vivo and in vitro environments [22]. In the body, nearly all tissue cells are supported by an ECM that comprises a complex, three-dimensional (3D), fibrous mesh network of collagen and elastic fibers integrated into a high hydrated, gel-like material containing proteoglycans, glycosaminoglycans, and glycoproteins. This complex system is responsible for the triggering of various biochemical and physical signals [23]. In practice, most cell culture studies are carried out using cells cultured as two-dimensional (2D) monolayers on hard plastic surfaces due to the convenience, ease, and high cell proliferation that these culture techniques provide. On the other hand, cell adaption to an artificial monolayer culture on an inflexible surface would lead to metabolic and functional alterations, resulting in behavior different from the in vivo environment [23]. Thus, research is focused on developing more controllable 3D cell culture matrices resembling, as much as possible, the in vivo conditions that are able to support cell growth, differentiation, and organization. Bearing in mind all the above, we can define 3D cell culture as the integration of cells into a hydrogel matrix in order to receive signals from the scaffold and surrounding cells [23,24]. This procedure initially necessitates a cellular suspension in a hydrogel precursor solution, and then entrapment through a gel initiation reaction that leads to the formation of covalently- or noncovalently-linked molecules [25,26].

A great number of synthetic and natural polymers can be used for cell entrapment gelled into hydrophilic matrices under mild conditions with minimal loss of viability [27]. The properties of the gel, i.e., either hydrophobic either hydrophilic, and its porosity, can be regulated. The entrapment of cells constitutes one of the most widely used methods for living cell immobilization within spherical beads of calcium alginate. This method is considered successful due to the fact that immobilization is a simple, quick and cost-effective technique, and is usually performed under very mild conditions [28].

Three-dimensional (3D) printing is a cheap additive layer manufacturing technology that is able to create sophisticated and complex-shaped bodies in a short time, especially for rapid prototyping engineering applications [29–31]. In many cases, 3D printing technology uses polymers, depending on the specific characteristics that are required for the microfabricated devices (e.g., polycarbonate (PC), PLA, nylon, polymethyl methacrylate (PMMA), polystyrene, and polyethylene terephthalate glycol (PETG)) [32–35]. Depending on the area of interest, 3D printing technology is increasing rapidly and is extensively used in many fields, such as bio-printing, medical devices, the automotive industry, soft sensors and actuators, space, art and jewelry, education, and tissue printing [36,37]. PETG constitutes a copolymer known for its biocompatibility, chemical resistance, recyclability, and transparency [38]. It can be used in different applications in the food and medical industry, with an acceptable flammability rating [39]. Unfortunately, it presents low resistance to ultraviolet (UV) light and performs weakly against frictional contact and scratching.

The aim of this study is to develop a proof-of-concept bioelectrical profiling assay to study the reaction of various cancer cell lines exposed to a common anticancer drug as a function of the cell population density. For comparison purposes, bioelectrical impedance-based measurements were taken on both untreated immobilized cells and on cells treated with 5-FU. Thus, two gold-plated (Au) electrodes were embedded in a 3D-printed PETG well for impedance measurements on four cancer cell lines (SK-N-SH, HEK293, HeLa and MCF-7) immobilized in calcium alginate matrix. Cell cultures were realized in three population densities tested with various frequencies. In this way, a more detailed application of the bioelectrical analysis on an in vitro system for monitoring different responses between various cancer cells (control and treated with 5-FU) was possible.

2. Materials and Methods

2.1. Theory

Bioimpedance, as a passive electrical property, is described as the capability of biological tissue to impede electric current. Bioimpedance measurements detect the response to electrical activity (potential or current). Bioimpedance is a complex quantity, mainly determined by the resistance (R) of the total amount of body water and by the capacitance of the cell membrane [40]. Electrical Impedance (Z) is defined by the ratio of the voltage (V) to the current (I), and is quite similar to resistance. The basic difference is that impedance extends to the frequency domain, and thus, is used in AC circuits, while resistance mainly refers to DC applications. The equation for the calculation of electrical impedance is:

$$Z = V/I \tag{1}$$

The determination of electrical impedance requires not only the application of an alternating current across a biological tissue, but also the measurement of the consequential differential voltage of the tissue sample, described in the following equations:

$$I(\omega) = I_0 \cdot \cos(\omega t + \theta) \tag{2}$$

$$V(\omega) = V_0 \cdot \sin(\omega t + \psi) \tag{3}$$

where I_0 and V_0 represent the amplitude θ and ψ the phase of the current and voltage signal, respectively. Taking into account that both I_0 and V_0 are calculated at the same angular frequency, i.e., $\omega = 2\pi f$, electrical impedance can be described as $Z(\omega) = (V(\omega))/(I(\omega))$ [41]. The expression of Z (Equation (1)) as a complex function can be used either as the modulus of the absolute value and the phase shift, or as the real part, R, representing resistance, and the imaginary part, X, representing capacitance and inductance, respectively [42]. In the case of direct current (DC) application, the imaginary part would be zero. The inverse of the impedance is called admittance (Y) and describes the current flow.

Impedance and admittance constitute AC parameters, and both are frequency dependent. The EIS measurement procedure involves the characterization of the complex impedance over a wide range of frequencies, as shown in Equation (3):

$$Z(\omega) = |Z|(\cos \varphi + j \sin \varphi) = R + j X \tag{4}$$

2.2. Cell Culture

SK-N-SH neuroblastoma cells (ATCC® HTB-11™) were cultured under standard conditions (37 °C, 5% CO_2) in 90% Minimum Essential Medium (MEM) (Eagle) with Earle's balanced salt solution (BSS) (Biowest, Nuaillé, France) and fetal bovine serum (FBS) (Thermo Fisher Scientific, Waltham, MA, USA) to a final concentration of 10%, 2 mM l-glutamine, 1.5 g/L sodium bicarbonate, 0.1 mM non-essential amino acids, 1 U μg^{-1} antibiotics (penicillin/streptomycin), and 1.0 mM sodium pyruvate (Biowest, Nuaillé, France). HEK293 (ATCC® CRL-1573™), HeLa (ATCC® CRM-CCL-2™) and MCF-7 (ATCC® HTB-22™) cell lines were grown in Dulbecco's Modified Eagle Medium (Biochrom Gmbh, Berlin, Germany), supplemented with 10% Fetal Bovine Serum (Thermo Fisher Scientific, Waltham, MA, USA), 2 mM l-glutamine, 0.5 mM sodium pyruvate, and 1% Penicillin-Streptomycin (Biowest, Nuaillé, France) in T-75 flasks (Sarstedt AG & Co. KG, Nümbrecht, Germany). Subcultivation was done in a 1:10 ratio. Cells were detached from culture flasks by treatment with trypsin-EDTA for 3–10 min. After detachment, they were resuspended in the culture medium to inactivate any remaining trypsin activity. After centrifugation for 5 min (1000 rpm), they were resuspended in the medium at concentrations of 10^6, 2×10^6, and 4×10^6 cells/mL.

2.3. Cells Preparation/Immobilization

Cell immobilization was performed in calcium alginate as an immobilization matrix. Briefly, sodium alginate in 1.5% concentration, sterilized by autoclave (121 °C, 20 min), was mixed with 5×10^4, 10^5, and 2×10^5 cells to a 0.75% final concentration and poured together in the well. Then, a 1% $CaCl_2$ gelling solution was added for 10 s for cross-linking, and washed with phosphate buffered saline (PBS). After washing, the calcium alginate scaffolds with the cells were incubated in the culture medium. The next day, the culture medium was removed and replaced with 1% FBS medium and 1% FBS with different concentrations 5-FU (Sigma-Aldrich Chemie GmbH, Taufkirchen, Germany).

2.4. Cell Viability Assay

Cell viability was evaluated by a 3-(4, 5-dimethylthiazol-2-yl)-2,5-diphenyltetrazolium bromide (MTT) colorimetric assay [43] with 5-FU as the positive control. The concentration of 5-FU for each cell line, indicated in Table 1, was selected based on previously published data [44–48]. These concentrations gave at least 30% inhibition in cell proliferation after 24 h incubation. The next day, the cells were treated with 0.5 mg/mL MTT (Duchefa Biochemie, Haarlem, The Netherlands) and incubated with dye for 3 h. After incubation, the medium was removed and the cell containing alginate scaffolds were solubilized with 0.1 M ethylenediaminetetraacetic acid per well. Cell morphology observations were performed with an inverted microscope (ZEISS Axio Vert.A1, Carl Zeiss Microscopy, LLC, White Plains, NY, USA), and pictures were processed using the ZEN lite software. The optical absorbance was measured with a PowerWave240 plate reader (BioTek, Winooski, VT, USA) at 560 nm. The experiment was repeated independently three times for each treatment, and the mean results were expressed as the average OD for each treatment. All values were presented as means ± SD, and significance testing in the comparisons was based on Student's T-tests for pairs, as they did not follow a normal distribution. The Student's T-test gives the probability that the difference between the two means is caused by chance. p values < 0.05 were considered to be statistically significant.

Table 1. 5-FU concentrations added to each cell culture.

	SK-N-SH	HEK293	HeLa	MCF-7
5-FU concentration (μM)	7.5	20	150	150

2.5. Experimental Setup

In order to perform the bioelectrical measurements, a specifically designed electrode-based system was fabricated. Figure 1a illustrates the experimental setup of the system used for the measurement procedure. More specifically, two gold-coated (Au) electrodes were placed vertically into a custom-made, transparent, 3D-printed PETG well that was designed to be used as a cell cultivation vessel. The application of 3D printing technology to the assembly of culture systems can provide an appropriate environment for cell growth which is able to mimic physiological and realistic cell phenotypes [49]. The well model was designed using the 123D Design software (Autodesk, San Rafael, CA, USA). A Cel Robox 3D printer device was utilized for the printing procedure, applying the fused deposition modeling technique (FDM). By this method, the filament that passes through the heated print head is laid on a construction platform in a layer-by-layer fashion until the object's form is complete [50]. The nozzle diameter of the print head was 0.4 mm and the printing temperature for PETG was 190 °C [38]. The PETG filament was obtained by the Formfutura BV (Nijmegen, The Netherlands); the filament diameter was 1.75 ± 0.05 mm. After the printing process, the wells were sterilized with 70% (v/v) ethanol for 10 min, and then dried for 2 h under a sterile hood. The electrodes were connected to the handheld LCR meter U1733C (Figure 1b) from Keysight Technologies (Santa Rosa, CA, USA); the instrument is able to measure at three frequencies (1 KHz, 10 KHz, and 100 KHz) for the direct extraction of impedance magnitude of the sample tested. For impedance measurements, a voltage of 0.74 V_{rms} ± 50 mV_{rms} was applied via the two terminals to the gold-coated electrodes. The best sampling rate of the instrument was 1 Hz (one measurement per sec); each measurement lasted one minute; thus, the total values obtained for each run were 60, with a measurement frequency of 1 Hz. All data were normalized values, presented as the mean of the absolute (ABS) value of the control (plain cell culture medium) minus the absolute value of cells, for both cases (treated or untreated with 5-FU ± SD), as shown in Equation (5). Significance testing in comparisons was based on Student's T-tests for pairs, and p values < 0.05 were considered to be statistically significant.

$$\text{Normalized value} = \text{mean}\ (|\text{control-cell value}|) \tag{5}$$

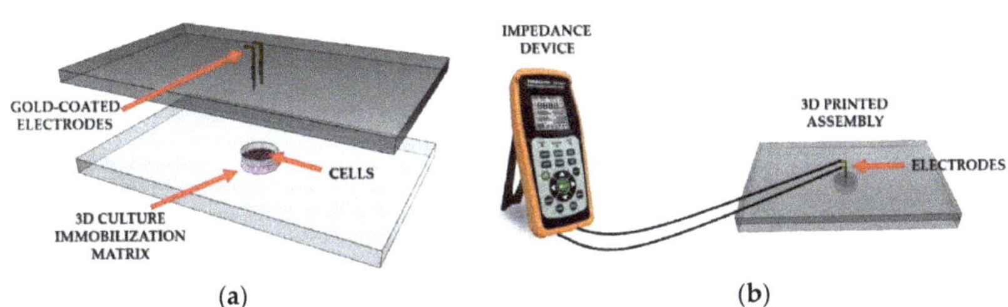

Figure 1. Experimental setup. (**a**) Representation of the cell chamber filled with 3D cell immobilization matrix; (**b**) Connection of the LCR meter to the 3D printed well.

3. Results

In this study, we evaluated the applicability of impedance measurements for the bioelectric profiling of different cancer cell types treated with substance-selected anticancer agents. More specifically, four cancer cell lines were immobilized in calcium alginate and cultured in different cell population densities (50,000, 100,000, and 200,000/100 μL). Then, 5-fluorouracil (5-FU) was applied, as it constitutes one of the most common cancer therapeutic drugs. In each case, three frequencies were tested: 1 KHz, 10 KHz, and 100 KHz.

3.1. Cell Proliferation

In order to ensure that calcium alginate was a proper immobilization matrix for the cancer cell culture, we assessed cellular viability with the MTT uptake assay. Cells were cultured in the matrix for 24 h (with and without treatment with 5-FU), and the proliferation was determined microscopically and photometrically after MTT application. Figures 2–5 depict the microscopic observations for three different populations of the four cell lines immobilized in calcium alginate after incubation with MTT.

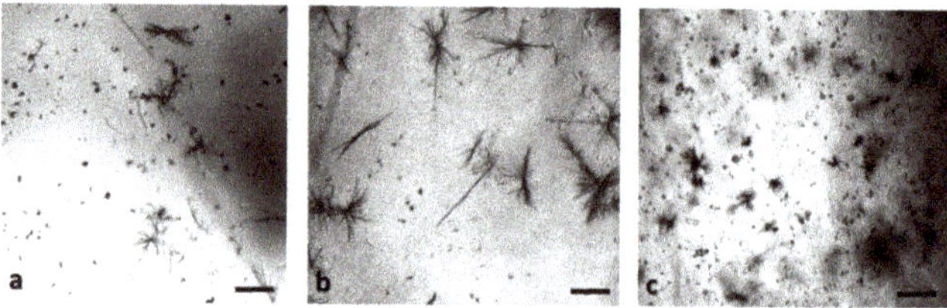

Figure 2. Panoramic view of SK-N-SH immobilized cells in 3D matrix after treatment with MTT for 24 h, showing the viability in three different population densities: (**a**) 50,000 cells; (**b**) 100,000 cells; and (**c**) 200,000 cells. Scale bars = 50 μm.

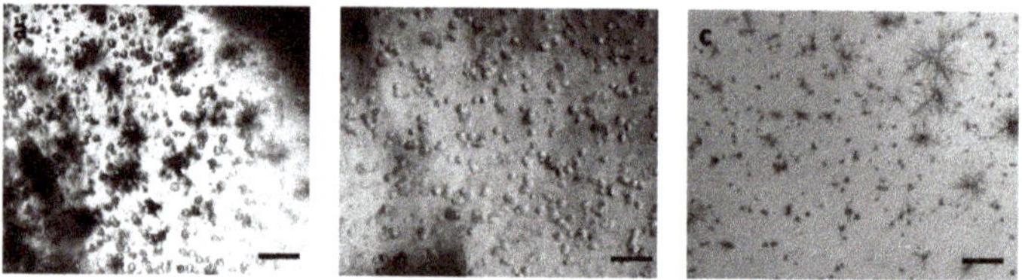

Figure 3. Panoramic view of HEK293 immobilized cells in 3D matrix after treatment with MTT for 24 h showing the viability in three different population densities: (**a**) 50,000 cells; (**b**) 100,000 cells; and (**c**) 200,000 cells. Scale bars = 50 μm.

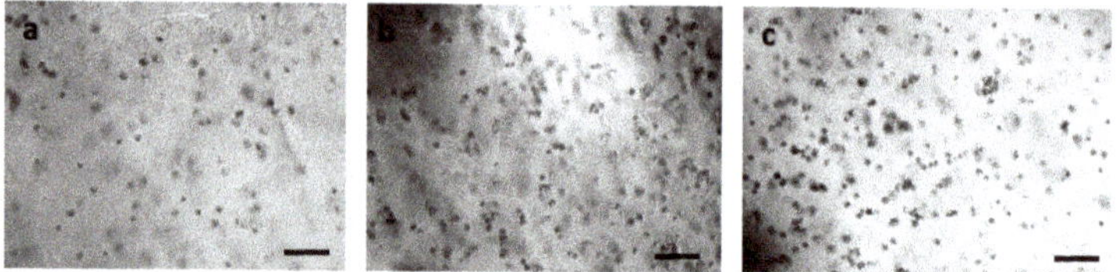

Figure 4. Panoramic view of HeLa immobilized cells in 3D matrix after treatment with MTT for 24 h showing the viability in three different population densities: (**a**) 50,000 cells; (**b**) 100,000 cells; and (**c**) 200,000 cells/100 μL. Scale bars = 50 μm.

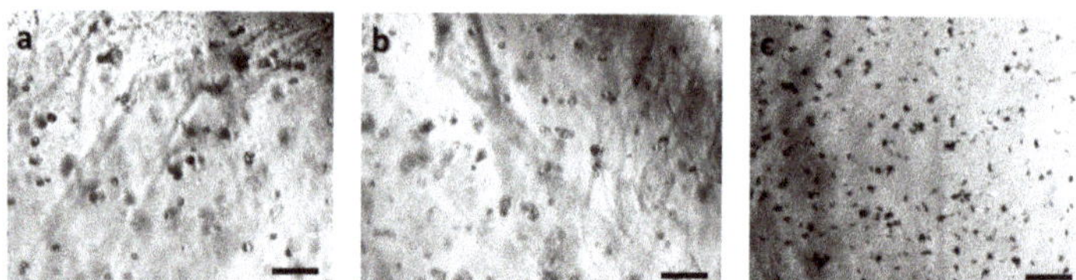

Figure 5. Panoramic view of MCF-7 immobilized cells in 3D matrix after treatment with MTT for 24 h showing the viability in three different population densities: (**a**) 50,000 cells; (**b**) 100,000 cells; and (**c**) 200,000 cells/100 μL. Scale bars = 50 μm.

Viable cells were dyed purple using the yellow formazan (MTT) by intracellular NAD(P)H-oxidoreductases [43]. We can see that cellular proliferation is affected neither by the immobilization matrix, nor by the increase in the cell population density. Contrary to this observation, the results from the photometric MTT determination presented in Figures 6–9 showed an increase in the absorbance as cell number population densities increase, whereas the addition of 5-FU led to a significant reduction in cell viability (see Table 2) in almost all cell lines. Cell population alterations in the neuroblastoma SK-N-SH cell line (see Figure 6) appear to have a limited impact in MTT absorbance for both cell cases, i.e., treated with 5-FU and untreated. On the other hand, in the case of the remaining cell lines (Figures 7–9), we observed an increase in absorbance proportional to the cell number.

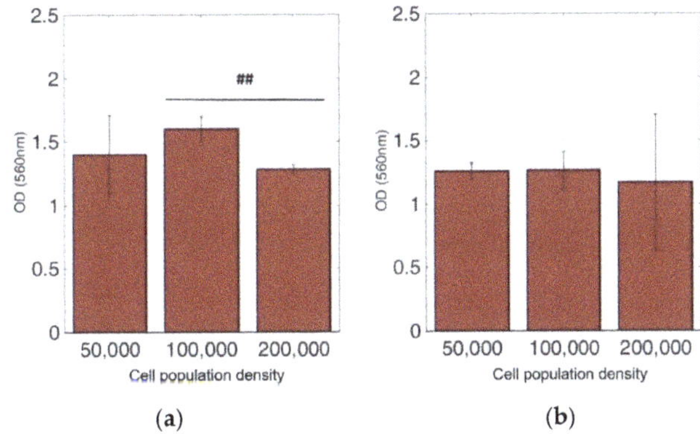

Figure 6. Cellular viability of SK-N-SH cells immobilized in 3D matrix after treatment with MTT for 24 h showing the viability in three different population densities (50,000, 100,000, and 200,000 cells/100 µL) ± STD: (**a**) untreated cells (control); (**b**) cells treated with 5-FU. ## < 0.01 significantly different from 100,000 cells/100 µL.

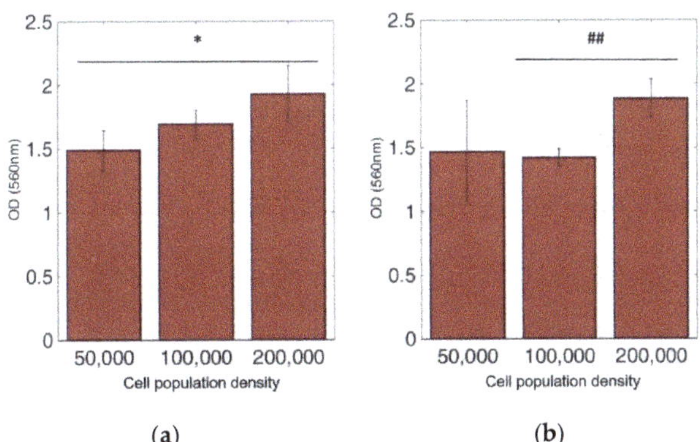

Figure 7. Cellular viability of HEK293 cells immobilized in 3D matrix after treatment with MTT for 24 h showing the viability in three different population densities (50,000, 100,000, and 200,000 cells/100 µL) ± STD: (**a**) untreated cells (control); (**b**) cells treated with 5-FU. * < 0.05 significantly different from 50,000 cells, ## < 0.01 significantly different from 100,000 cells/100 µL.

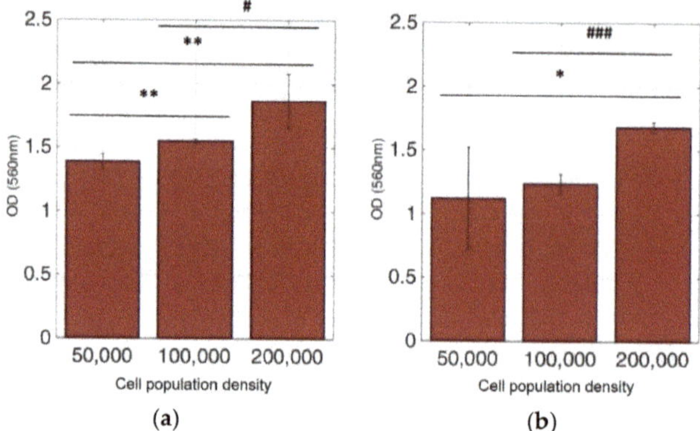

Figure 8. Cellular viability of HeLa cells immobilized in 3D matrix after treatment with MTT for 24 h showing the viability in three different population densities (50,000, 100,000, and 200,000 cells/100 µL) ± STD: (**a**) untreated cells (control); (**b**) cells treated with 5-FU. * < 0.05, ** < 0.01 significantly different from 50,000 cells, # < 0.05, ### < 0.001 significantly different from 100,000 cells/100 µL.

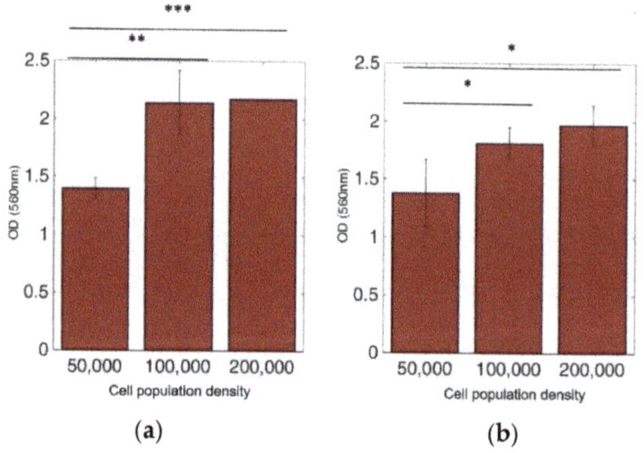

Figure 9. Cellular viability of MCF-7 cells immobilized in 3D matrix after treatment with MTT for 24 h showing the viability in three different population densities (50,000, 100,000, and 200,000 cells/100 µL) ± STD: (**a**) untreated cells (control); (**b**) cells treated with 5-FU. * < 0.05, ** < 0.01, *** < 0.001 significantly different from 50,000 cells.

Table 2. Significant differences (Student's T-test) between cell populations before and after treatment with 5-FU. **< 0.01, *** < 0.001.

	50,000 Cells	100,000 Cells	200,000 Cells
SK-N-SH	-	**	-
HEK293	-	**	-
HeLa	-	***	-
MCF-7	-	-	-

For further analysis, all cell line combinations were compared using a Student's T-test in each population density, with or without the addition of 5-FU. As shown in Table 3, in the 50,000 cell/100 µL population density, treatment with 5-FU did not significantly affect MTT uptake. However, it seems that the other two population densities contributed to differential viability results, i.e., with or without treatment with 5-FU.

Table 3. Significant differences (Student's T-test) in cell viability between different cell lines X cell population densities before and after treatment with 5-FU. * < 0.05, **< 0.01, *** < 0.001.

	Cells			Cells Treated with 5-FU		
	50,000 Cells	100,000 Cells	200,000 Cells	50,000 Cells	100,000 Cells	200,000 Cells
SK-N-SH–HEK293	-	-	*	-	-	*
SK-N-SH–HeLa	-	-	**	-	-	-
SK-N-SH–MCF-7	-	*	***	-	***	*
HEK293–HeLa	-	*	-	-	**	*
HEK293–MCF-7	-	*	-	-	**	-
HeLa–MCF-7	-	**	*	-	**	*

3.2. *Comparative Bioelectrical Profiling Results among Different Immobilized Cell Lines*

This experimental approach refers to the analysis of bioelectrical impedance-based measurements on various cancer cell types in different population densities. Calcium alginate was once again chosen as the 3D immobilization matrix used for each cancer cell culture. Figures 10–13 depict the absolute values of the differences between the mean blank values and the mean impedance values for three population densities tested for each cell line chosen in three different frequencies (1 KHz, 10 KHz, and 100 KHz).

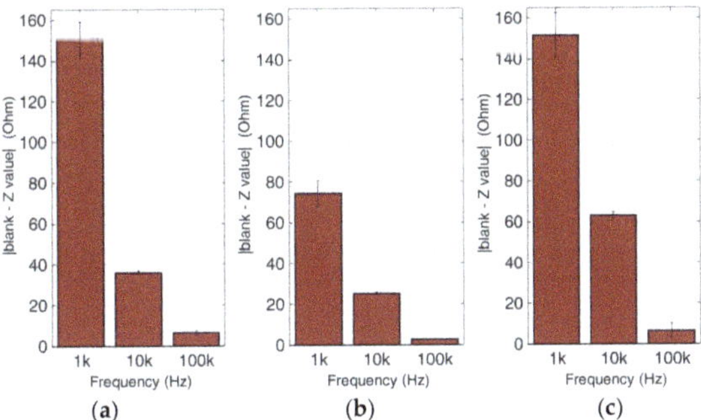

Figure 10. Normalized values of the mean impedance magnitude for untreated (control) immobilized SK-N-SH cancer cell lines tested at three frequencies (1 KHz, 10 KHz, 100 KHz) for three different population densities ± STD: (**a**) 50,000 cells; (**b**) 100,000 cells and (**c**) 200,000 cells/100 μL.

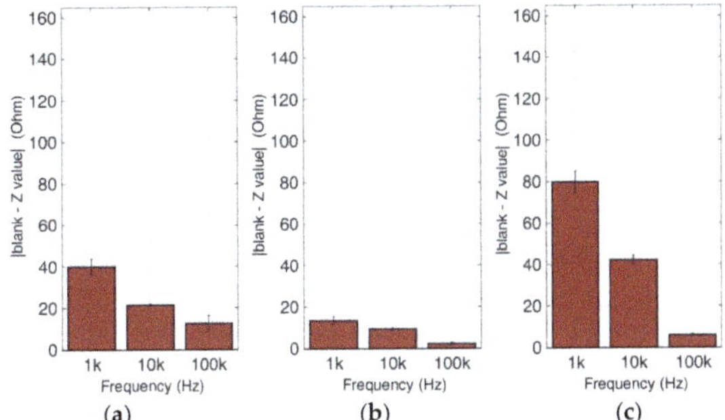

Figure 11. Normalized values of the mean impedance magnitude for untreated (control) immobilized HEK293 cancer cell lines tested at three frequencies (1 KHz, 10 KHz, 100 KHz) for three different population densities ± STD: (**a**) 50,000 cells; (**b**) 100,000 cells and (**c**) 200,000 cells/100 μL.

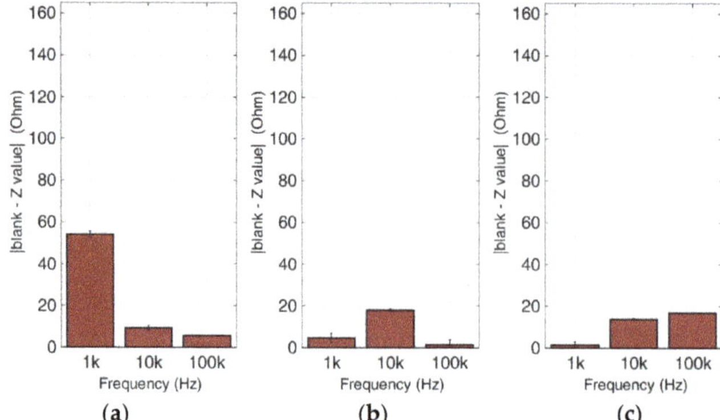

Figure 12. Normalized values of the mean impedance magnitude for untreated (control) immobilized HeLa cancer cell lines tested at three frequencies (1 KHz, 10 KHz, 100 KHz) for three different population densities ± STD: (**a**) 50,000 cells; (**b**) 100,000 cells and (**c**) 200,000 cells/100 μL.

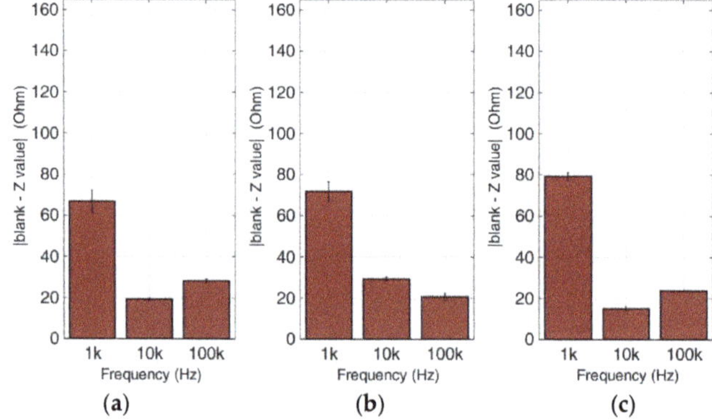

Figure 13. Normalized values of the mean impedance magnitude for untreated (control) immobilized MCF-7 cancer cell lines tested at three frequencies (1 KHz, 10 KHz, 100 KHz) for three different population densities: (**a**) 50,000 cells; (**b**) 100,000 cells and (**c**) 200,000 cells/100 μL.

As indicated by many studies, and also in our case, a frequency-dependent impedance response is observed. Each cell line x population density combination corresponds to a different pattern in the impedance magnitude measured in each frequency. Thus, in neuroblastoma SK-N-SH cells (see Figure 10), we can observe that the normalized impedance value drops when we move from 50,000 to 100,000 cells, and then increases at 200,000 cells/100 μL at each frequency. A similar pattern is observed in the case of HEK293 cells, but as shown in Figure 11, at 200,000 cells/100 μL population density, the impedance value is higher than the respective impedance in 50,000 cells. However, HeLa and MCF-7 cell cultures (Figures 12 and 13) appear to have a totally different behavior. More specifically, at a lower frequency (1 KHz), HeLa cells initially demonstrated a high impedance value (54, 23 Ohm), and as population density increased, a dramatic decrease was observed (1, 60 Ohm). In contrast, MCF-7 cells demonstrated the opposite trend, starting with 66, 72 Ohm at 50,000 cells and ending up at 79, 23 Ohm at 200,000 cells/100 μL. Furthermore, at the other two frequencies, we observed differential fluctuations for both cell lines. In other words, each cell line x population density combination was characterized by its own unique impedance behavior fingerprint.

3.3. Comparative Bioelectrical Profiling Results among Different Immobilized Cell Lines Treated with 5-FU

At this experimental stage, the evaluation of 5-FU applied to the previous immobilized cancer cell cultures was implemented to investigate the effects of this widely-used anticancer medicine through impedance measurements at specific frequencies. Figures 14–17 depict the normalized impedance values after the subtraction of the mean blank values from the mean impedance values in three population densities tested for each cell line chosen in three different frequencies, with (yellow bars) or without (blue bars) 5-FU. The statistical significances after pair comparisons for all combinations of the cell populations are represented in Tables 4–7.

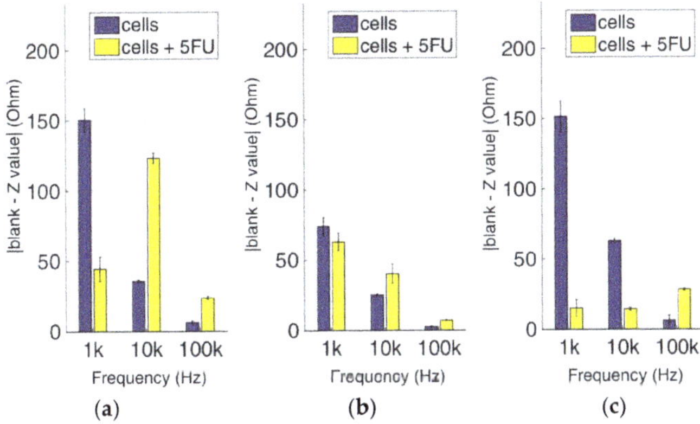

Figure 14. Normalized values of the mean impedance magnitude for control immobilized SK-N-SH cells (blue bars) and immobilized SK-N-SH cells treated with 5-FU (yellow bars), tested at three different cell population densities ± STD: (**a**) 50,000; (**b**) 100,000 and (**c**) 200,000/100 μL for three different frequencies (1 KHz, 10 KHz, and 100 KHz).

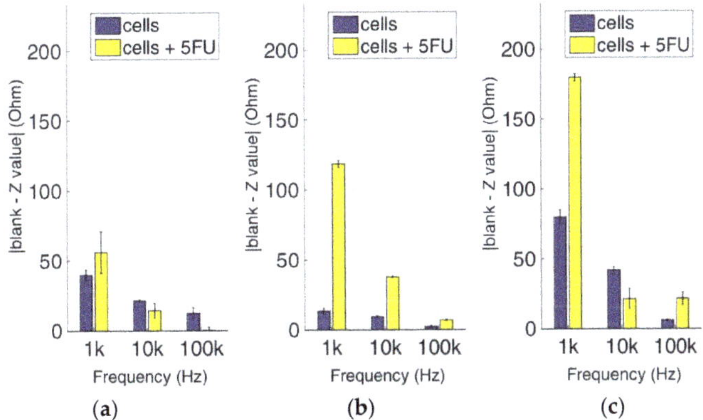

Figure 15. Normalized values of the mean impedance magnitude for control immobilized HEK293 cells (blue bars) and immobilized HEK293 cells treated with 5-FU (yellow bars), tested at three different cell population densities ± STD: (**a**) 50,000; (**b**) 100,000 and (**c**) 200,000/100 μL for three different frequencies (1 KHz, 10 KHz, and 100 KHz).

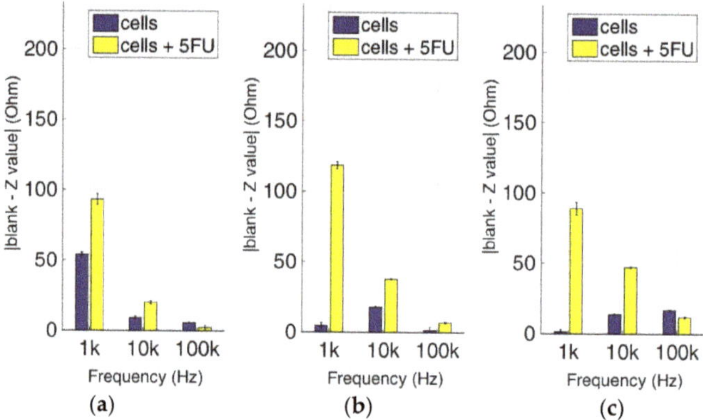

Figure 16. Normalized values of the mean impedance magnitude for control immobilized HeLa cells (blue bars) and immobilized HeLa cells treated with 5-FU (yellow bars), tested at three different cell population densities ± STD: (**a**) 50,000; (**b**) 100,000 and (**c**) 200,000/100 μL for three different frequencies (1 KHz, 10 KHz, and 100 KHz).

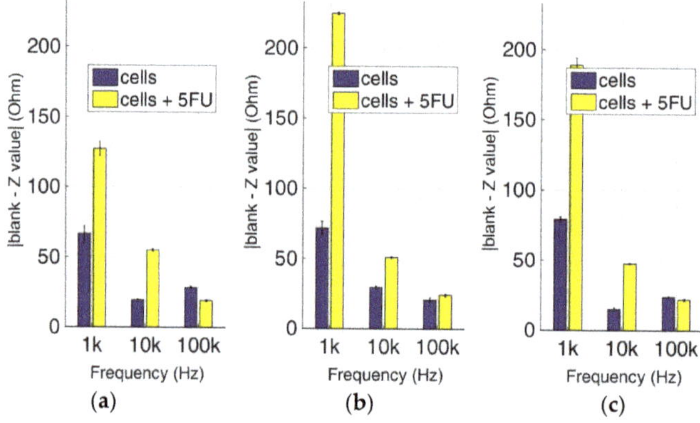

Figure 17. Normalized values of the mean impedance magnitude for control immobilized MCF-7 cells (blue bars) and immobilized MCF-7 cells treated with 5-FU (yellow bars), tested at three different cell population densities ± STD: (**a**) 50,000; (**b**) 100,000 and (**c**) 200,000/100 μL for three different frequencies (1 KHz, 10 KHz, and 100 KHz).

Table 4. Significant differences (Student's T-test) between population densities for the SK-N-SH cell line before and after treatment with 5-FU. **< 0.01, *** < 0.001.

	Cells			**Cells Treated with 5-FU**		
	1 KHz	**10 KHz**	**100 KHz**	**1 KHz**	**10 KHz**	**100 KHz**
50,000–100,000	***	***	**	**	***	***
50,000–200,000	-	***	-	***	***	***
100,000–200,000	***	***	-	***	**	***

Table 5. Significant differences (Student's T-test) between population densities for HEK293 cell line before and after treatment with 5-FU. * < 0.05, **< 0.01, *** < 0.001.

	Cells			Cells Treated with 5-FU		
	1 KHz	10 KHz	100 KHz	1 KHz	10 KHz	100 KHz
50,000–100,000	***	***	**	***	***	**
50,000–200,000	***	***	*	***	-	**
100,000–200,000	***	***	***	***	*	**

Table 6. Significant differences (Student's T-test) between population densities for HeLa cell line before and after treatment with 5-FU. * < 0.05, **< 0.01, *** < 0.001.

	Cells			Cells Treated with 5-FU		
	1 KHz	10 KHz	100 KHz	1 KHz	10 KHz	100 KHz
50,000–100,000	***	***	*	***	***	**
50,000–200,000	***	**	***	-	***	***
100,000–200,000	*	***	***	***	***	***

Table 7. Significant differences (Student's T-test) between population densities for MCF-7 cell line before and after treatment with 5-FU. * < 0.05, **< 0.01, *** < 0.001.

	Cells			Cells Treated with 5-FU		
	1 KHz	10 KHz	100 KHz	1 KHz	10 KHz	100 KHz
50,000–100,000	-	***	**	***	***	***
50,000–200,000	**	***	***	***	***	***
100,000–200,000	*	***	*	***	***	**

When the anticancer agent 5-fluorouracil was applied for 24 h, we observed that in most cases (18 out of 27), 5-FU treatment gave higher impedance normalized values in comparison to cells with no treatment. The SK-N-SH cell line (see Figure 14) followed mostly the opposite pattern when 5-FU was applied compared to the rest of the cell lines, especially at a frequency of 1 KHz. In the case of 200,000 cells/100 μL population density, the cell impedance magnitude followed a downward trend, as opposed to cells under 5-FU exposure, that gave higher values in the frequency range of 1–100 KHz.

Figure 15 summarizes the results for the HEK293 cell line. Similar to the aforementioned observations, impedance significantly dropped with an increase in frequency in all population densities, and treatment with 5-FU led to higher normalized values (see Table 5). An exception can be observed at the frequencies of 10 KHz and 100 KHz in 50,000 cell population density, and also at the frequency of 10 KHz in 200,000 cells/100 μL population density, where the values obtained from control cells were higher than the respective values with 5-FU.

As depicted in Figure 16, HeLa cells treated with 5-FU follow a frequency-dependent, downward motif for every population density. On the other hand, untreated cells do not show the same pattern, since a downward trend is observed at 50,000 cells/100 μL, followed by an upward tendency at 200,000 cells/100 μL, and general non-linear behavior at 100,000 cells/100 μL. Once again, in almost all cases, the response of the cells treated with 5-FU is significantly higher in comparison with control cells (Table 6). The only exception is observed in 50,000 and 200,000 cells/100 μL population densities at 100 KHz. The latter frequency gave low impedance values for both the treated and untreated cell populations.

In the MCF-7 cell line (Figure 17), we observed particularly high impedance values when 5-FU was applied for every population density tested when compared with untreated cells, especially at a frequency of 1 KHz. These values significantly increased with population density at the same frequency (Table 7). The impedance values of cells not treated with 5-FU depicted low variations between different cell population densities. A general observation is that as frequency magnitude increases, the normalized impedance values decrease.

4. Discussion

Up to now, the development of cancer diagnostics was primarily controlled by direct tumor tissue biopsies for either pathologic and/or histologic analyses. Novel advanced molecular biology techniques such as next-generation DNA sequencing and genomics bioinformatics analysis represent examples of the transition from traditional microscopy of tissue samples to molecular genomics for cancer diagnosis. In combination with remarkable advances in drug development and efficiency, these exciting new trends have contributed to the transition to the era of personalized cancer diagnosis and therapy. Therefore, these advances in cancer management and treatment have incentivized the pursuit of avant-garde, non-invasive approaches for accurate detection and monitoring.

For this purpose, in this study, we investigated the differences between the electrical properties of different in vitro 3D cancer cell cultures such as cervical, breast, and kidney tumor models via impedance evaluation at various frequencies. In general, several variances are observed in cell activities such as morphology, proliferation, and gene and protein expression due to the additional dimensionality of 3D culture, compared to the 2D planar culture [51]. Hydrogels such as alginate (SA), collagen (COL), fibrin, and agarose (AG) have attracted the most attention as promising matrices for bioinks because of their innate biocompatibility, low cytotoxicity, and high water content, mimicking natural ECM [52–54]. SA, which offers fast gelling in the presence of Ca^{2+} or other divalent cations, was frequently used as a bioink for cells to be easily and quickly encapsulated, and for interlayer adhesion during the layer-by-layer printing process [55–57]. Studies have indicated that alginate and agarose bioinks support the development of hyaline-like cartilage tissues, whereas GelMA- and PEGMA-based bioinks favor the development of fibrocartilaginous tissues [58,59]. The main disadvantage of chitosan is that it provides poor mechanical integrity to the tissue, making the 3D bioprinted structures brittle and delicate. Finally, the major pitfall with fibrin use is the fast and irreversible gelation at body temperature, which makes its bioprinting difficult [60].

An optical inverted microscope was used for the observation of the differences in cell morphology between the four different cell lines after conducting the MTT colorimetric assay [43]. Similar observations were reported for studies on epithelial cancer cells in 3D culture [61–63]. Our results indicated that the immobilization matrix did not affect cell viability. Furthermore, the photometric MTT determination revealed an increase in cellular proliferation relative to the cell population density. On the other hand, when the anticancer agent 5-fluorouracil (5-FU) was added to the cell medium, viability was significantly reduced, suggesting that the 3D immobilization matrix does not influence the influx of the compound in the alginate hydrogel.

After completing the biochemical cytotoxic assays, a bioelectrical analysis by means of impedance measurements was performed on each cancer cell line coupled with its corresponding electrodes in the 3D alginate matrix. The measurement protocol was divided into two cases, based on different aspects of the cell cultures we wanted to investigate. In the first stage, impedance measurements were recorded for different population densities (50,000, 100,000, and 200,000 cells/100 ul) of the aforementioned cell lines in 3D cultures in plain medium. In the second step, the respective measurements were evaluated for 3D cultures after 24 h application of the anticancer compound, 5-FU.

We used the resulting characteristic features to determine contrasts between distinctive cell types by means of normalized impedance magnitudes. The method has been particularly effective in discriminating not only cells of different tissue origin, but also the cytotoxic impact of the anticancer compound.

The method can provide useful information as the assay provides information about the response of cells in specific frequency values, giving us the opportunity to utilize it as a putative cancer diagnostic technique. For further research, this methodology could efficiently be expanded to additional cell lines, i.e., cancerous or normal ones. Similar studies have been conducted on skin [64], breast [65], esophagus [66], and cervical cancer cells [6].

The discrimination method in this study describes a measurement procedure and data processing which is quite easy to handle, although its application for cancer diagnoses or assessments of efficient chemotherapy treatment would require samples that can, in reality, be obtained only by invasive means. Nevertheless, in most cases a definite diagnosis of a malignant tumor depends upon assaying a real tissue sample. That said, the results of the present study demonstrate that even a small number of cells (obtainable through, e.g., liquid biopsy) could be bioelectrically profiled in order to determine their behavior with respect to their susceptibility to selected anticancer agents. While this has not been investigated in the framework of this study, it may be possible to achieve an even higher sensitivity in terms of detecting a low number of cells (i.e., much fewer than 50,000/mL, in the order of a few hundreds or thousands), especially if a considerably wider range of impedance frequencies is used (see also below).

Specific frequency values in the range of 1 KHz–100 KHz were selected to study the electrical properties of the different cell types. Particularly high impedance normalized values were observed at 1 KHz for both treatments, with or without 5-FU, in almost all cell types. At this frequency value, the contribution of cell structure becomes relevant, as the cell-hydrogel interface is influenced by the properties of either the plain cell culture medium and/or the anticancer drug. Hence, cell structure might play an important role in detection sensitivity. For cancerous tissues, a decrease of impedance at frequencies higher than 1 kHz was observed, while in the frequency level up to approximately 100 KHz, these changes were not visible [67]. In addition, it has been reported that the impedance of abnormal tissues, such as breast cancer tissue [68], has lower values compared to those for healthy tissues [69].

Moreover, supplementary investigations could evaluate the prospect of performing a complementary assay on in vivo samples. Our consideration focuses on the fact that these results can begin to highlight the diversities in the electrical behavior of normal and cancerous 3D cell cultures during the whole measurement procedure. The assessment of more heterogeneous models with additional characteristic features, such as resistance or capacitance, could help the enhancement of the system's resolution capacity. Although our methodology has succeeded in effectively determining various cancer cell types, our next goal is to adjust it to specify the existence of cancer cells in cocultures with normal ones in a single well containing a set of suitably-placed electrodes.

5. Conclusions

In this study, we present the proof-of-concept development of a complementary, non-invasive cell analysis method to assess the responses of 3D cultured cancer cell lines derived from various tissues following treatment with cytotoxic concentrations of the well-studied anticancer compound, 5-FU, using bioelectricalal evaluation. A key-feature of the cell-based bioelectrical sensor was the 3D printing of a PETG well, assembled with two gold-coated electrodes, perpendicularly wall-mounted at the bottom, allowing for impedance measurements. The evaluation of our cell bio-system configuration showed good efficacy towards different cell type determination with adequate sensitivity. In order to obtain higher levels of selectivity, and sensitivity based on very low cell populations (up to single-cell analysis), it is necessary to fabricate a different electrode configuration (e.g., screen-printed) which is suitable for single-cell impedance spectroscopy [70–72]. As a next step, we plan to conduct more detailed impedance spectral analyses by using a considerably wider continuous frequency range, rather than the discrete values that were used in the present approach, as well as to investigate the effect of different cell immobilization agents on the bioelectric profiling process.

Author Contributions: Conceptualization—G.K. and S.K.; Data curation—G.P. and S.M.; Formal analysis—G.P. and S.M.; Investigation—G.P. and S.M.; Methodology—G.K. and S.K.; Resources—S.M.; Validation—G.P.; Visualization—G.P.

Funding: This work is supported by the H2020-EU.1.3.3. program of the topic MSCA-RISE-2017 entitled "NanoFEED: Nanostructured carriers for improved cattle feed" (Grant agreement ID: 778098).

Conflicts of Interest: The authors declare no conflict of interest.

References

1. Lindsey, A.T.; Islami, F.; Siegel, R.L.; Ward, M.E.; Jemal, A. Global Cancer in Women: Burden and Trends. *Cancer Epidemiol. Biomark. Prev.* **2017**, *26*, 444–457.
2. Cruz, C.S.D.; Tanoue, L.T.; Matthay, R.A. Lung Cancer: Epidemiology, Etiology, and Prevention. *Clin. Chest Med.* **2011**, *32*, 605–644. [CrossRef] [PubMed]
3. Izetbegovic, S.; Alajbegovic, J.; Mutevelic, A.; Pasagic, A.; Masic, I. Prevention of Diseases in Gynecology. *Int. J. Prev. Med.* **2013**, *4*, 1347–1358. [PubMed]
4. Nigel, S.K.; Alok, A.K.; Nicole, M.K.; Kari, B.; Agnes, Y.Y.L.; Juan, I.A.; Sandra, L.W.; Edward, P.B.; Christopher, R.F.; Charles, W.F.; et al. Venous Thromboembolism Prophylaxis and Treatment in Patients With Cancer: ASCO Clinical Practice Guideline Update. *J. Clin. Oncol.* **2019**. [CrossRef]
5. Montagnana, M.; Lippi, G. Cancer diagnostics: Current concepts and future perspectives. *Ann. Transl. Med.* **2017**, *5*, 268. [CrossRef] [PubMed]
6. Hong, J.L.; Lan, K.C.; Jang, L.S. Electrical characteristics analysis of various cancer cells using a microfluidic device based on single-cell impedance measurement. *Sens. Actuators B Chem.* **2012**, *173*, 927–934. [CrossRef]
7. Arruebo, M.; Vilaboa, N.; Sáez-Gutierrez, B.; Lambea, J.; Tres, A.; Valladares, M.; González-Fernández, Á. Assessment of the Evolution of Cancer Treatment Therapies. *Cancers* **2011**, *3*, 3279–3330. [CrossRef]
8. Park, S.Y.; Choi, J.B.; Kim, S. Measurement of cell-substrate impedance and characterization of cancer cell growth kinetics with mathematical model. *Int. J. Precis. Eng. Manuf.* **2015**, *16*, 1859–1866. [CrossRef]
9. Liu, D.; Li, X.; Chen, X.; Sun, Y.; Tang, A.; Li, Z.; Zheng, J.; Shi, M. Neural regulation of drug resistance in cancer treatment. *BBA-Rev. Cancer* **2019**, *1871*, 20–28. [CrossRef]
10. Yadav, A.K.; Agarwal, A.; Rai, G.; Mishra, P.; Jain, S.; Mishra, A.K.; Agrawal, H.; Agrawal, G.P. Development and characterization of hyaluronic acid decorated PLGA nanoparticles for delivery of 5-fluorouracil. *Drug Deliv.* **2010**, *17*, 561–572. [CrossRef]
11. Paivana, G.; Apostolou, T.; Kaltsas, G.; Kintzios, S. Study of the dopamine effect into cell solutions by impedance analysis. In Proceedings of the Bio-Medical Instrumentation and related Engineering and Physical Sciences, Athens, Greece, 12–13 October 2017.
12. Paivana, G.; Apostolou, T.; Mavrikou, S.; Barmpakos, D.; Kaltsas, G.; Kintzios, S. Impedance Study of Dopamine Effects after Application on 2D and 3D Neuroblastoma Cell Cultures Developed on a 3D-Printed Well. *Chemosensors* **2019**, *7*, 6. [CrossRef]
13. Kilic, T.; Navaee, F.; Stradolini, F.; Renaud, P.; Carrara, S. Organs-on-chip monitoring: Sensors and other strategies. *Microphysiol. Syst.* **2018**, *2*, 5. [CrossRef]
14. Qiao, G.; Duan, W.; Sinclair, A.; Wang, W.W. Electrical properties of breast cancer cells from impedance measurement of cell suspensions. *J. Phys. Conf. Ser.* **2010**, *224*, 012081. [CrossRef]
15. Chaoshi, R.; Huiyan, W.; Yuan, A.; Hong, S.; Guojing, L. *Development of Electrical Bioimpedance Technology in the Future*; IEEE: Piscataway, NJ, USA, 2002.
16. Altinagac, E.; Taskin, S.; Kizil, H. Single Cell Array Impedance Analysis for Cell Detection and Classification in a Microfluidic Device. In Proceedings of the 10th International Joint Conference on Biomedical Engineering Systems and Technologies (BIOSTEC 2017), Porto, Portugal, 21–23 February 2017; pp. 49–53.
17. Hanbin, M.; Wallbank, R.W.R.; Chaji, R.; Li, J.; Suzuki, Y.; Jiggins, C.; Nathan, A. An impedance-based integrated biosensor for suspended DNA characterization. *Sci. Rep.* **2013**, *3*, 2730.
18. Bera, T.K. Bioelectrical Impedance Methods for Noninvasive Health Monitoring: A Review. *J. Med. Eng.* **2014**, *2014*, 381251. [CrossRef]
19. Asphahani, F.; Zhang, M. Cellular impedance biosensors for drug screening and toxin detection. *Analyst* **2007**, *132*, 835–841. [CrossRef]
20. Wang, M.-H.; Kao, M.-F.; Jang, L.-S. Single HeLa and MCF-7 cell measurement using minimized impedance spectroscopy and microfluidic device. *Rev. Sci. Instrum.* **2011**, *82*, 064302. [CrossRef]
21. Heileman, K.; Daoud, J.; Tabrizian, M. Dielectric spectroscopy as a viable biosensing tool for cell and tissue characterization and analysis. *Biosens. Bioelectron.* **2013**, *49*, 348–359. [CrossRef]
22. Pampaloni, F.; Reynaud, E.G.; Stelzer, E.H. The third dimension bridges the gap between cell culture and live tissue. *Nat. Rev. Mol. Cell Biol.* **2007**, *8*, 839. [CrossRef]

23. Lee, J.; Cuddihy, M.J.; Kotov, N.A. Three-dimensional cell culture matrices: State of the art. *Tissue Eng. Part B Rev.* **2008**, *14*, 61. [CrossRef]
24. Prestwich, G.D. Simplifying the extracellular matrix for 3-D cell culture and tissue engineering: A pragmatic approach. *J. Cell. Biochem.* **2007**, *101*, 1370. [CrossRef] [PubMed]
25. Drury, J.L.; Mooney, D.J. Hydrogels for tissue engineering: Scaffold design variables and applications. *Biomaterials* **2003**, *24*, 4337. [CrossRef]
26. Place, E.S.; George, J.H.; Williams, C.K.; Stevens, M.M. Synthetic polymer scaffolds for tissue engineering. *Chem. Soc. Rev.* **2009**, *38*, 1139. [CrossRef] [PubMed]
27. Nagaoka, T.; Shiigi, H.; Tokonami, S.; Saimatsu, K. Entrapment of Whole Cell Bacteria into Conducting Polymers. *J. Flow Inject. Anal.* **2012**, *29*, 7–10.
28. Willaert, R. Cell Immobilization and Its Applications in Biotechnology: Current Trends and Future Prospects. *Cell Immobil. Appl. Biotechnol.* **2011**, *12*, 313.
29. Kitson, P.J.; Rosnes, M.H.; Sans, V.; Dragone, V.; Cronin, L. Configurable 3D-printed millifluidic and microfluidic 'lab on a chip' reactionware devices. *Lab Chip* **2012**, *12*, 183267–183271. [CrossRef] [PubMed]
30. Waldbaur, A.; Rapp, H.; Lange, K.; Rapp, B.E. Let there be chip-Towards towards rapid prototyping of microfluidic devices: One-step manufacturing processes. *Anal. Methods* **2011**, *3*, 2681–2716. [CrossRef]
31. Munshi, A.S.; Martin, R.S. Microchip-Based Electrochemical Detection using a 3-D Printed Wall-Jet Electrode Device. *Analyst* **2016**, *141*, 862–869. [CrossRef]
32. Kwon, K.W.; Park, M.C.; Lee, S.H.; Kim, S.M.; Suh, K.Y.; Kim, P. Soft lithography for microfluidics: A review. *Biochip. J.* **2008**, *2*, 1–11.
33. Liu, K.; Fan, Z.H. Thermoplastic microfluidic devices and their applications in protein and DNA analysis. *Analyst* **2011**, *136*, 1288–1297. [CrossRef]
34. Becker, H.; Nevitt, M.; Gray, B.L. Selecting and designing with the right thermoplastic polymer for your microfluidic chip: A close look into cyclo-olefin polymer. *Microfluid. BioMEMS Med. Microsyst. XI* **2013**, *8615*, 86150F.
35. Bhattacharyya, A.; Klapperich, C.M. Thermoplastic microfluidic device for on-chip purification of nucleic acids for disposable diagnostics. *Anal. Chem.* **2006**, *78*, 788–792. [CrossRef] [PubMed]
36. Sharma, S.; Goel, S.A. Three-Dimensional Printing and its Future in Medical World. *J. Med. Res. Innov.* **2018**, *3*, e000141. [CrossRef]
37. Garcia, J.; Yang, Z.; Mongrain, R.; Leask, R.L.; Lachapelle, K. 3D printing materials and their use in medical education: A review of current technology and trends for the future. *BMJ Simul. Technol. Enhanc. Learn.* **2018**, *4*, 27–40. [CrossRef] [PubMed]
38. Yuan, L.; Ding, S.; Wen, C. Additive manufacturing technology for porous metal implant applications and triple minimal surface structures: A review. *Bioact. Mater.* **2019**, *4*, 56–70. [CrossRef]
39. Otero, J.J.; Vijverman, A.; Mommaerts, M.Y. Use of fused deposit modeling for additive manufacturing in hospital facilities: European certification directives. *J. Cranio-Maxillofac. Surg.* **2017**, *45*, 1542–1546. [CrossRef]
40. Grimnes, S.; Martinsen, G. *Bioimpedance & Bioelectricity, Basics*, 3rd ed.; Acaademic Press: Cambridge, MA, USA, 2014.
41. Barsoukov, E.; Macdonald, J.R. *Impedance Spectroscopy Theory, Experiment, and Applications*, 2nd ed.; Wiley Interscience Publication: Hoboken, NJ, USA, 2005.
42. Ramírez-Chavarría, R.G.; Sanchez-Perez, C.; Matatagui, D.; Qureshi, N.; Perez-García, A.; Hernandez-Ruíz, J. Ex-vivo biological tissue differentiation by the Distribution of Relaxation Times method applied to Electrical Impedance Spectroscopy. *Electrochim. Acta* **2018**, *276*, 214–222. [CrossRef]
43. Mosmann, T. Rapid colorimetric assay for cellular growth and survival: Application to proliferation and cytotoxicity assays. *J. Immunol. Methods* **1983**, *65*, 55–63. [CrossRef]
44. Li, J.; Wang, X.; Hou, J.; Huang, Y.; Zhang, Y.; Xu, W. Enhanced anticancer activity of 5-FU in combination with Bestatin: Evidence in human tumor-derived cell lines and an H22 tumor-bearing mouse. *Drug Dev. Ther.* **2015**, *9*, 45–52. [CrossRef]
45. Maney, V.; Singh, M. The Synergism of Platinum-Gold Bimetallic Nanoconjugates Enhances 5-Fluorouracil Delivery In Vitro. *Pharmaceutics* **2019**, *11*, 439. [CrossRef]
46. Hemaiswarya, S.; Doble, M. Combination of phenylpropanoids with 5-fluorouracil as anti-cancer agents against human cervical cancer (HeLa) cell line. *Phytomedicine* **2013**, *20*, 151–158. [CrossRef] [PubMed]

47. Kazi, J.; Mukhopadhyay, R.; Sen, R.; Jha, T.; Ganguly, S.; Debnath, M.C. Design of 5-fluorouracil (5-FU) loaded, folate conjugated peptide linked nanoparticles, a potential new drug carrier for selective targeting of tumor cells. *Med. Chem. Commun.* **2019**, *10*, 559–572. [CrossRef] [PubMed]
48. Pasquier, E.; Ciccolini, J.; Carré, M.; Giacometti, S.; Fanciullino, R.; Pouchy, C.; Montero, M.; Serdjebi, C.; Kavallaris, M.; André, N.V. Propranolol potentiates the anti-angiogenic effects and anti-tumor efficacy of chemotherapy agents: Implication in breast cancer treatment. *Oncotarget* **2011**, *2*, 797–809. [CrossRef] [PubMed]
49. Antoni, D.; Burckel, H.; Josset, E.; Noel, G. Three-dimensional cell culture: A breakthrough in vivo. *Int. J. Mol. Sci.* **2015**, *16*, 5517–5527. [CrossRef]
50. JWeisman, A.; Nicholson, J.C.; Tappa, K.; Jammalamadaka, U.; Wilson, C.G.; Mills, D.K. Antibiotic and chemotherapeutic enhanced three-dimensional printer filaments and constructs for biomedical applications. *Int. J. Nanomed.* **2015**, *10*, 357–370.
51. Schwartz, M.A.; Chen, C.S. Deconstructing dimensionality. *Science* **2013**, *339*, 402–404. [CrossRef]
52. Chung, J.H.Y.; Naficy, S.; Yue, Z.; Kapsa, R.; Quigley, A.; Moulton, S.E.; Wallace, G.G. Bio-ink properties and printability for extrusion printing living cells. *Biomater. Sci.* **2013**, *1*, 763–773. [CrossRef]
53. Frampton, J.P.; Hynd, M.R.; Shuler, M.L.; Shain, W. Fabrication and optimization of alginate hydrogel constructs for use in 3D neural cell culture. *Biomed. Mater.* **2011**, *6*, 015002. [CrossRef]
54. Guillotin, B.; Guillemot, F. Cell patterning technologies for organotypic tissue fabrication. *Trends Biotechnol.* **2011**, *29*, 183–190. [CrossRef]
55. Axpe, E.; Oyen, M.L. Applications of Alginate-Based Bioinks in 3D Bioprinting. *Int. J. Mol. Sci.* **2016**, *17*, 1976. [CrossRef]
56. Yeo, M.G.; Kim, G.H. A cell-printing approach for obtaining hASC-laden scaffolds by using a collagen/polyphenol bioink. *Biofabrication* **2017**, *9*, 025004. [CrossRef] [PubMed]
57. Kim, Y.B.; Lee, H.; Kim, G.H. Strategy to Achieve Highly Porous/Biocompatible Macroscale Cell Blocks, Using a Collagen/Genipin-bioink and an Optimal 3D Printing Process. *ACS Appl. Mater. Interfaces* **2016**, *8*, 32230–32240. [CrossRef] [PubMed]
58. Daly, A.C.; Critchley, S.E.; Rencsok, E.M.; Kelly, D.J. A comparison of different bioinks for 3D bioprinting of fibrocartilage and hyaline cartilage. *Biofabrication* **2016**, *8*, 045002. [CrossRef] [PubMed]
59. Zhang, Y.; Yu, Y.; Ozbolat, I.T. Direct bioprinting of vessel-like tubular microfluidic channels. *J. Nanotechnol. Eng. Med.* **2013**, *4*, 020902. [CrossRef]
60. Murphy, S.V.; Skardal, A.; Atala, A. Evaluation of hydrogels for bio-printing applications. *J. Biomed. Mater. Res.* **2013**, *101*, 272–284. [CrossRef] [PubMed]
61. Weaver, V.M.; Petersen, O.W.; Wang, F.; Larabell, C.; Briand, P.; Damsky, C.; Bissell, M.J. Reversion of the malignant phenotype of human breast cells in three-dimensional culture and in vivo by integrin blocking antibodies. *J. Cell Biol.* **1997**, *137*, 231–245. [CrossRef]
62. Lee, G.Y.; Kenny, P.A.; Lee, E.H.; Bissell, M.J. Three-dimensional culture models of normal and malignant breast epithelial cells. *Nat. Methods* **2007**, *4*, 359–365. [CrossRef]
63. Zhao, Y.; Yao, R.; Ouyang, L.; Ding, H.; Zhang, T.; Zhang, K.; Cheng, S.; Sun, W. Three-dimensional printing of Hela cells for cervical tumor model in vitro. *Biofabrication* **2014**, *6*, 035001. [CrossRef]
64. Moqadam, S.M.; Grewal, P.K.; Haeri, Z.; Ingledew, P.A.; Kohli, K.; Golnaraghi, F. Cancer detection based on electrical impedance spectroscopy: A clinical study. *J. Electr. Bioimpedance* **2018**, *9*, 17–23. [CrossRef]
65. Huerta-Nuñez, L.F.E.; Gutierrez-Iglesias, G.; Martinez-Cuazitl, A.; Mata-Miranda, M.M.; Alvarez-Jiménez, V.D.; Sánchez-Monroy, V.; Golberg, A.; González-Díaz, C.A. A biosensor capable of identifying low quantities of breast cancer cells by electrical impedance spectroscopy. *Sci. Rep.* **2019**, *9*, 6419. [CrossRef]
66. Wang, H.-C.; Nguyen, N.-V.; Lin, R.-Y.; Jen, C.-P. Characterizing Esophageal Cancerous Cells at Different Stages Using the Dielectrophoretic Impedance Measurement Method in a Microchip. *Sensors* **2017**, *17*, 1053. [CrossRef] [PubMed]
67. Fraczek, M.; Krecicki, T.; Moron, Z.; Krzywaznia, A.; Ociepka, J.; Rucki, Z.; Szczepanik, Z. Measurements of electrical impedance of biomedical objects. *Acta Bioeng. Biomech.* **2016**, *18*, 11–17. [PubMed]
68. Pethig, R. Dielectric properties of body tissues. *Clin. Phys. Physiol. Meas.* **1987**, *8*, 5–12. [CrossRef] [PubMed]
69. Wtorek, J. *Techniki Elektroimpedancyjne w Medycynie*; Wydawnictwo PG: Gdańsk, Poland, 2002.

70. Charwat, V.; Joksch, M.; Sticker, D.; Purtscher, M.; Rothbauer, M.; Ertl, P. Monitoring cellular stress responses using integrated high-frequency impedance spectroscopy and time-resolved ELISA. *Analyst* **2014**, *139*, 5271–5282. [CrossRef]
71. Holmes, D.; Pettigrew, D.; Reccius, C.H.; Gwyer, J.D.; van Berkel, C.; Holloway, J.; Davies, D.E.; Morgan, H. Leukocyte analysis and differentiation using high speed microfluidic single cell impedance cytometry. *Lab Chip* **2009**, *9*, 2881–2889. [CrossRef]
72. Malleo, D.; Nevill, J.T.; Lee, L.P.; Morgan, H. Continuous differential impedance spectroscopy of single cells. *Microfluid. Nanofluidics* **2010**, *9*, 191–198. [CrossRef]

Article

Attack Graph Modeling for Implantable Pacemaker

Mariam Ibrahim [1,*], Ahmad Alsheikh [2] and Aseel Matar [3]

1 Department of Mechatronics Engineering, German Jordanian University, Amman 11180, Jordan
2 Department of Natural Sciences and Industrial Engineering, Deggendorf Institute of Technology, 94469 Deggendorf, Germany; a.alsheikh@gju.edu.jo
3 Department of Biomedical Engineering, German Jordanian University, Amman 11180, Jordan; As.Matar@gju.edu.jo
* Correspondence: mariam.wajdi@gju.edu.jo

Received: 10 January 2020; Accepted: 17 February 2020; Published: 19 February 2020

Abstract: Remote health monitoring systems are used to audit implantable medical devices or patients' health in a non-clinical setting. These systems are prone to cyberattacks exploiting their critical vulnerabilities. Thus, threatening patients' health and confidentiality. In this paper, a pacemaker automatic remote monitoring system (PARMS) is modeled using architecture analysis and design language (AADL), formally characterized, and checked using the JKind model checker tool. The generated attack graph is visualized using the Graphviz tool, and classifies security breaches through the violation of the security features of significance. The developed attack graph showed the essentiality of setting up appropriate security measures in PARMS.

Keywords: pacemaker; threat modeling; internet of things (IoT) medical devices; vulnerabilities

1. Introduction

Implantable therapeutic tools are becoming progressively interdependent through the internet of things (IoT) in order to audit vital signs and improve patients' quality of life. Yet, the IoT imposes major vulnerabilities with such interconnection, and any disturbance could cause significant destruction or life-impeding demands [1,2]. An adversary may construct various attacks to jeopardize both IoT implantable therapeutic equipment and networks [3]. Table 1 illustrates some recent cyberattack incidents that occurred in the medical field. Thus, it is not easy to design and protect medical devices that are able to cope with equipment failures and connectivity and operating systems faults [4]. Security and privacy concerns should also be considered, such as identification, data integrity, confidentiality, authentication, and user and service privacy [5]. A recent survey [6] studied over one hundred medical tools to consider their protection worries with a focus on reported cyberattacks including tampering, sniffing, and unauthorized access. The survey also studied available mitigation methods to handle these worries.

Table 1. Cyberattack incidents in the medical field.

Date	Country	Name	Description
August 2011	United States	Medtronic insulin-delivery system	Hacked the insulin pump and completely disabled it [7]
2008	United States	Cardiac defibrillator	Hacked cardiac defibrillatorto change the device's settings, ordering it to deliver a shock, and disabling it [7]
2017	United Kingdom	16 United Kingdom hospitals	Freezing systems and encrypting files [8]

Table 1. *Cont.*

Date	Country	Name	Description
2014	United States	Boston Children's Hospital	Caused the hospital network to lose internet access using distributed Denial of Service (DoS) attack [9]
January, 2015	United States	Anthem	Breached a database with 80 million customers records [10]
July, 2018	England	National Health Service (NHS)	A data breach caused the NHS to share confidential health data of 150,000 patients [11]
June, 2018	Singapore	SingHealth	The data of 1.5 million patients were stolen [12]
2019	United States	NeuroSky 156 brain–computer interface application	Victims' brain wave data were stolen [13]

Attack graphs provide a viewable technique to determine risks within interoperable systems. The actions needed to conduct an attack can be identified utilizing this technique. The identification of attacks helps engineers to establish defensive actions in order to eliminate the execution of an attack [14]. For instance, a method is presented by [15] for indicating the best placement of a collection of IoT tools within an institution using a traditional attack graph which is augmented to consider the substantial placement of IoT tools and their connectivity effectiveness.

Attack graphs can also help forensic investigators to identify many possible attack paths. An empirical study is provided by [16] on the growth of using data gathered by smartphone tools (developed to correlate a therapeutic tool) as digital clue in legal cases. A report is included about evidence which is possibly helpful in a digital forensics inspection.

A digital inspection system is proposed by [17] for the examination of fatal attack scenarios on cardiac implantable medical devices (IMDs). The system reports the identification and regeneration of possible attack scenarios that result in a patient's death. An approach of three stages is proposed, along with a collection of approaches to use in every stage. In the first stage, the approach aids determining the reason for a death based on the therapeutic conclusions gathered by the IMD. Second, the approach follows the entries and system logs gathered from the IMD under consideration, which determine the critical actions associated with distant access and construction. The technique aims to collect the possible attack scenarios that could achieve similar impact in the gathered log proof, as if they had been conducted. A library of threats and a model checking established algorithm are utilized to conduct the automatic reformation which is made in forward chaining. The third stage of the approach correlates the generated scenarios, identifies the most persuasive composite of medical and vocational scenarios, and confirms the presence of abnormal attitude in the chosen composite that caused a patient's death.

The main contribution of this work manifests an approach for developing attack graphs for the pacemaker automatic remote monitoring system (PARMS). This demands a general specification of system model (design and communications, units, resources, protections, vulnerabilities, and attack instances), and exploration of the security concerns. The model and the security properties are encoded using architecture analysis and design language (AADL) [18] and verified using JKind checker embedded software [19]. The developed attack graph contains the attack scenarios causing system compromise through gaining ability to alter the settings of the home monitoring device. Thus, controlling the wireless pacemaker and jeopardizing the patient's life. The resulting graph is visualized utilizing Graphviz [20]. The rest of this paper is organized as follows: Section 1.1 reviews the relevant work. Section 2 presents the modeling process of the pacemaker automatic remote monitoring system (PARMS). Section 3 illustrates attack graph construction and visualization for the PARMS. Section 4 recaps and discusses some forthcoming work.

1.1. Related Work

Different papers were investigated in the literature for modelling attack graphs for medical devices. A model-based system, a safety and security co-engineering (MB3SE) technique, and a correlated toolchain for the implementation of medical equipment was proposed by [21]. The toolchain included architecture modelling and safety and cyber-security risk analysis tools. Explanations for security concerns of 5G networks aiding electronic healthcare applications were presented by [22]. The explanations incorporated knowledge graph development, automated attack and protection technologies, and a security testbed.

An approach is presented by [23] for developing attack trees for IMDs which receive two inputs: functional workflow and a hazard study of the IMD in consideration. A process-modeling software is utilized to illustrate the IMD system as it is arranged, booted up, and managed by the caregiver. Hazards can be identified as system states that are built-in unprotected for the user. Hazard study requires determining system states that will ultimately cause critical harm to the patient.

Threat modeling is examined in medical cyber physical systems (MCPS) by [24]. This includes the roles of stakeholders and system components, trust models, threat models, and threat analysis. An abstract architecture is also sketched for an MCPS to demonstrate various threat modeling options.

A methodology has been developed by [2] for generating attack trees for patient controlled analgesia (PCA-IMD). This process contains four steps: (1) process modeling, (2) fault tree analysis (FTA), (3) attack tree generation, and (4) quantification. First, the user of the PCA-IMD takes a depiction of the workflow of the PCA-IMD and constructs a process-modeling design for it. Once the process model is constructed, the IMD user establishes the distinct hazards that can happen as a result of running the system, leading to extra infusion.

Two internal activities are studied by [25], involving the utilization of Universal Serial Bus (USB) drives and Compact Disc Read-Only Memory (CD-ROM) as the entrance methods leading to data loss in the healthcare firm surroundings. The generated augmented threat trees show the vulnerabilities abused, the actions required to abuse them, and the fingerprint implemented by the attackers' functionalities. A Markov models set is developed by [26] for a healthcare IoT foundation, that enables the consideration of the particularity of clients' machines, connectivity, advancement of data stream, and protection and security worries of these elements.

The modeling and study of cyberattacks utilizing a multimodal graph technique is shown by [27]. This work illustrates how cyber actions, parties, targets, and networks that gathered them can be modeled using a multimodal graph, such that multiple graphs of distinct modalities are connected to show the features of the attack.

A framework is presented by [28] for modeling and assessing security of the IoT which incorporates preprocessing, security model generation using a hierarchical attack representation model (HARM), conception and repository, security study, and transformations and updates. In the scheme, an IoT, security model generators, and an evaluator are implemented.

The authors of [29] investigated whether the ideas of model checking and attack tree refinement correspond to using an IoT healthcare illustrative example. The extension by model checking and the enclosing of attack trees into the Isabelle internal scheme permitted the investigation of this correspondence utilizing the analytical strict and automated proof assistance of Isabelle. Hence, reassessing the interpretation of state evolution in model checking and importing a variation that showed the attack sequences. This permitted the conversion of attack paths established by model checking into the attack tree refinement procedure.

An attack graph-based study is presented by [4] of attacks on a certain interoperability surrounding to provide patient pain medication (PCA) among multiple levels of interoperability from simple data gathering to complete closed loop control. Explanations of the potential prevention methods are determined for every class of attack vectors. The work showed that security has a deep impact on the safety of medical device interoperability and the patients they are provided to.

Conceptual graphs are collected by [30] with Dung's disputation system that supplied convenient extensions for dependable selection procedures, all adapted to telemedicine in general and tele-expertise in particular. The work implemented the visual graph of attacks where distinct interpretation of the reasoning logic is adapted to verify the possible adequate arguments.

A systematic threat-modeling approach is proposed by [31] to investigate IMD security. The attack tree approach provided an overall and organized scheme of the strengths and weaknesses of the IMD system. The work showed a systematic method for conducting system-level security examination to incorporate various potential attack surfaces. The research done by [14] demonstrated attack graph modeling on hypothetical ambulatory medical equipment. The research examined specific attacks that jeopardized ambulatory equipment, like physical attacks and social engineering.

2. Pacemaker Automatic Remote Monitoring System (PARMS) Modeling

In this work, the pacemaker automatic remote monitoring system (PARMS) is modeled to illustrate how hacking into the pacemaker's system imposes life-threatening risks to patients. The model includes system topology, possible attack instances, and the system's formal description.

2.1. PARMS Topology

Figure 1 shows a modified pacemaker automatic remote monitoring system (PARMS) from [32]. The PARMS includes the following components:

- **Wireless Pacemaker**: This is a battery-powered implantable device that produces an excitatory wave at an appropriate site within the heart. The pacemaker initiates the electrical depolarization cycle of the heart at approximately 72 beats/min in the atrioventricular (AV) node to replace a malfunctioning AV node. The AV node is the electrical connecting point from the atria to the ventricles, which continues excitation beyond a partial or total heart block. Modern pacemakers can also store diagnostic data [33].

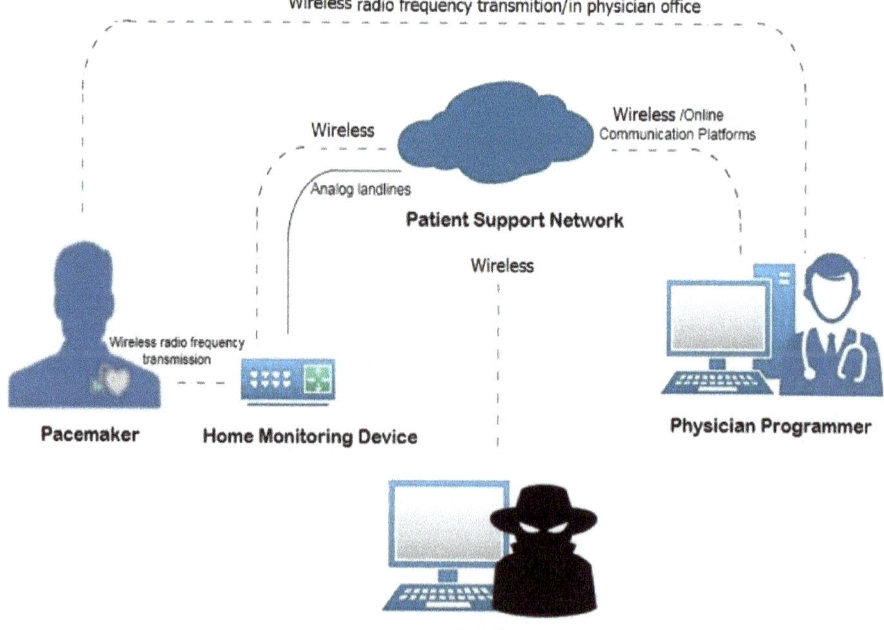

Figure 1. pacemaker automatic remote monitoring system (PARMS).

The pacemaker consists of a cardiac pulse generator, which is a device with a power source and electronic circuitry that produces the excitatory signal. The pulse generator contains a power source "battery" to power it. It also contains a programmable segment to evaluate the heart function and send the appropriate electric impulse signal from the pulse generator to the heart [33]. Other components of the pacemaker are the leads which conduct electrical signals from the pulse generator to the tissue and, in some pacemakers, also conduct signals from the tissue to the pulse generator.

The pacemaker has also an electrode which is located at the end of the lead which continuously measures the heart rate. The sensed signal is then amplified using a low noise amplifier. Next, the amplified signal is filtered using a second order low pass filter, and the result is an appropriate electrocardiogram (ECG) signal. This signal is then applied to a comparator which compares it to a defined threshold to detect the heartbeat event executed by the heart and generates a pulse with every heartbeat using a pulse generator [34]. The wireless pacemakers are embedded with a micro-antenna to enable wireless communication with the home monitoring device and the programmer [35].

- **Physician Programmer (PP)**: A computer with specific software and associated hardware modifications is used to program the pacemaker at the time of implantation. It is used to test pacemaker functionality in the follow-up visits to the physician's office by sending instructions to change therapy parameters, and reading battery status and heart rhythms. The physician programmer contains a transceiver which communicates with the pacemaker antenna via radio frequency (RF) [32,35].
- **Home Monitoring Device (HS)**: This is a transceiver placed in proximity to the patient and used to monitor the pacemaker by collecting the data that the pacemaker sends periodically at a set of frequencies (e.g., every two days or every week) using RF transmissions via the micro-antenna. The HS then sends data to the physician programmer via analog landlines or wireless data networks. The data incorporate heart rate, battery status, pacing lead impedance, episodes of arrhythmias, conveyed antitachycardia pacing, percent pacing, histogram, real-time and magnet electrograms (EGM), reserved EGMs, arrhythmia reviews, and mode switch period [35,36].
- **Patient Support Network (PN)**: This works on the principle of cloud computing, where the data transmitted from the home monitoring device can be stored into the network servers and can be uploaded onto a secure website that the physician may log in to, to review the data. [32,36].
- **Access Point (AP)**: This exists for outside internet communication. We assume the attacker is located at this point.

Communication among the PARMS' components can be summarized as follow [32]:

i. RF communication between the PP and the wireless pacemaker in order to program it and check its functionality.
ii. The PP and the PN are connected through online communication platforms: "cloud networking".
iii. RF communication between the HS and the wireless pacemaker to inquire different measurements related to the pacemaker such as the battery state and pulse rate.
iv. The communication between the HS and the PN involves sending measurements to the PN to be stored there and accessed later by the responsible PP. The PN also transmits updates from the network to the HS using analog landlines, a wireless data network, or a wireless Global System for Mobile communications (GSM).

To illustrate how hacking into the pacemaker's system presents life-threatening risks to patients, two vulnerabilities are identified within PARMS. These are the microprocessor commercial off-the-shelf (COTS) vulnerability, and the firmware update vulnerability due to loss of validation of the source of firmware updates [32]. These vulnerabilities can be exploited resulting in the following possible attack instances:

1. *Intelligent Gathering (IG)*: This is used for gathering information about IoT devices such as Internet Protocol (IP) addresses, and checking the type of firmware, as well as the existence of COTS.

2. *Social Engineering (SE)*: This attack is used to gain access and disclose information. It generally targets enterprises and organizations.
3. *Pivoting (PV)*: This is a standard technique used in penetration testing to navigate from machine to machine [37].
4. *Sniffing (S)*: This attack is used to steal or break off data by capturing the network traffic using a sniffer, for example to get login usernames and passwords that are sent by the HS [32].
5. *Man-in-the-Middle (MiM)*: This attack can occur when the attacker has access over the network connection. Thus, disclosing and manipulating the data flow between two parties.
6. *Phishing (PS)*: This is used to steal user data and allow the attacker to disguise as a trusted entity [38].
7. *Malware Injection (MA)*: The attacker can edit, copy, or install the code at the host. Thus, gaining a root access.
8. *SQL Injection (SQL)*: This attack is used to exploit the web application. Thus, allowing the attacker to gain unauthorized access to the PN database or retrieve information directly from it [39].

2.2. Formal Description of PARMS

The System can be formally described as follows:

1. The attacker is assumed to be located at (AP) and has a root privilege.
2. system components S; variable s ∈ {PP, PN, HS, AP} (static).
3. system connectivity, C ⊆ PP × PN, HS × PN; $c_{ij} = 1$ if component i is connected to component j (static).
4. System vulnerabilities V; Boolean $v_i = 1$ if vulnerability v ∈ {*COTS*, *firmware*} exists on i (static).
5. Set of possible attacks B; variable b ∈ {IG, SE, S, PS, PV, SQL, MiM, MA}.
6. Attack instances, AI ⊆ B × C; labeled b_{ij} ≡ attack b from source i to target j, b ∈ B.
7. Attacker level of privilege P on HS device; variable p_{HS} ∈ {none, root} (dynamic).
8. Attacker level of privilege PH on host i ∈ {PP, PN, AP}; variable ph_i ∈ {none, user, root} (dynamic).
9. Data identification D of component i; Boolean $d_i = 1$ if identification data about i gets collected by attacker (dynamic).
10. Confidential data disclosure K of component i; Boolean $k_i = 1$ if confidential data of component i get disclosed to attacker (dynamic).
11. Data alteration E of component i, Boolean $e_i = 1$ if attacker is able to edit the setting on component i (dynamic).
12. Attack instances pre-conditions:
    - $Pre(IG_{ij}) = (c_{ij} = 1) \wedge (ph_i = root) \wedge (p_{HS} = none)$
    - $Pre(SE_{ij}) = (c_{ij} = 1) \wedge (ph_i = root) \wedge (ph_j = none)$
    - $Pre(SE_{ij}) = (c_{ij} = 1) \wedge (ph_i = root) \wedge (ph_j = none)$
    - $Pre(S_{ij}) = (c_{ij} = 1) \wedge (p_{HS} = none) \wedge (d_{HS} = 1) \wedge (COTS_j = 1)$
    - $Pre(PS_{ij}) = (c_{ij} = 1) \wedge (ph_i = user) \wedge (d_j = 1) \wedge (firmware_j = 1)$
    - $Pre(PV_{ij}) = (c_{ij} = 1) \wedge (ph_i = user) \wedge (d_j = 1)$
    - $Pre(SQL_{ij}) = (c_{ij} = 1) \wedge (ph_i = root) \wedge (ph_j = user) \wedge (d_j = 1) \wedge (e_j = 1) \wedge (k_j = 1)$
    - $Pre(MIM_{ij}) = (c_{ij} = 1) \wedge (ph_i = user) \wedge (d_j = 1) \wedge (e_j = 0) \wedge (k_j = 1)$
    - $Pre(MA_{ij}) = (c_{ij} = 1) \wedge (ph_i = root) \wedge (p_{HS} = none) \wedge (d_j = 1) \wedge (e_j = 1) \wedge (k_j = 1) \wedge (COTS_j = 1 \vee firmware_j = 1)$
13. Attack instances post-conditions:
    - $Post(IG_{ij}) = (d_{HS} = 1) \wedge (p_{HS} = none)$
    - $Post\ (SE_{ij}) = (ph_j = user) \wedge (d_j = 1)$

- $Post\ (SE_{ij}) = (ph_j = user) \wedge (d_j = 1)$
- $Post\ (S_{ij}) = (ph_j = user) \wedge (d_j = 1) \wedge (k_j = 1)$
- $Post\ (PS_{ij}) = (ph_j = root) \wedge (d_j = 1) \wedge (e_j = 1) \wedge (k_j = 1)$
- $Post\ (PV_{ij}) = (ph_j = user) \wedge (d_j = 1) \wedge (k_j = 1)$
- $Post\ (SQL_{ij}) = (ph_j = root) \wedge (d_j = 1) \wedge (e_j = 1) \wedge (k_j = 1)$
- $Post\ (MIM_{ij}) = (ph_j = root) \wedge (d_j = 1) \wedge (e_j = 1) \wedge (k_j = 1)$
- $Post\ (MA_{ij}) = (p_{HS} = root) \wedge (d_{HS} = 1) \wedge (e_{HS} = 1) \wedge (k_{HS} = 1)$

14. Initial state: $ph_{AP} = root \wedge (\forall j \in \{PP, PN\}: ph_j = none \wedge p_{HS} = none \wedge (d_j = e_j = k_j = d_{HS} = e_{HS} = k_{HS} = 0))$. (Initially, the attacker has a root privilege on access point, no data identification, no confidential data disclosure, and no ability to alter the setting of HS).
15. Security property (φ): The attacker has no ability to edit the setting on the home monitoring device HS. Thus, jeopardizing patient's life. This property can be then described by a computational tree logic (CTL):

$$\varphi \equiv AG\ (e_{HS} = 0) \equiv AG\ (\neg\ (e_{HS} = 1))$$

3. Attack Graph Generation

Two software programs were utilized to conduct the cyberattack scenarios' generation and visualization, as shown in Figure 2. These tools are JKind model checker and Graphviz. JKind is a software tool that we used to conduct cyberattack scenarios against the PARMS [40]. The model checker keeps checking repeatedly if a given finite-state model of a system meets a given security property of importance. JKind is an infinite-state model checker for analyzing safety attributes of a system asserted in Lustre, a data flow synchronous terminology arranged for programming reactive systems like automatic control and auditing systems [41]. The JKind employs a back-end satisfiability modulo theories (SMT) solver to validate if a system model complies with a specific temporal logic property in every execution of the system. A wrong execution in which a property is not fulfilled is expressed as a counter example (CE) illustrating a sequence of attack instances (i.e., attack scenarios).

The PARMS depiction model of the parts and their interfaces and links is defined using architecture analysis and design language (AADL), within the open-source integrated development environment (Osate2). The AADL model is confined by assume guarantee reasoning environment (AGREE) annex plug-in in which the constants or variables are established locally. The AGREE plug-in translates the AADL+Annex models and properties to Lustre and communicates with JKind which verifies the system against the security property under study φ, and gives the result as a CE.

Considering the given security property φ, the goal of the attacker is to gain a root access on the home monitoring device (HS), and therefore gain the ability to alter the settings of the HS. Thus, imposing a life-threatening risk to patient. The JKind model checker generated the following counter example ($CE1: IG_{AP\text{-}HS} \rightarrow S_{HS\text{-}PN} \rightarrow MIM_{PN\text{-}PN} \rightarrow MA_{PN\text{-}HS}$) as a spreadsheet shown in Figure 3.

This attack sequence can be summarized as follows. Initially, the attacker has a root privilege on AP, an $IG_{AP\text{-}HS}$ attack is initialized to gather information about the HS (e.g., IP addresses). After the $IG_{AP\text{-}HS}$ attack an $S_{HS\text{-}PN}$ attack is launched between the HS and PN to get login username and password. This will allow the attacker to access the PN with user privilege therefore disclosing patient and HS information. Using the disclosed information, an $MIM_{PN\text{-}PN}$ attack is launched against the PN components to gain a higher privilege (root privilege). Using this privilege, an $MA_{PN\text{-}HS}$ attack is conducted exploiting a COTS vulnerability in the HS to gain a root access to it. By doing so, the attacker can alter the settings of the HS which will affect the wireless pacemaker and jeopardize the patient's life.

The generated counter example CE1 is encoded in disjunction with the property φ under study, that is $\varphi \vee CE1$. A new counterexample complies with: $\neg\ (\varphi \vee CE1) = \neg\ \varphi \wedge \neg\ CE1$, i.e., a counter example of φ distinct from $CE1$. This produces a new counter example ($CE2: SE_{AP\text{-}PN} \rightarrow PV_{PN\text{-}PN} \rightarrow MIM_{PN\text{-}PN} \rightarrow MA_{PN\text{-}HS}$). By continuing this process, three CEs were found, producing the complete attack scenarios (attack graph).

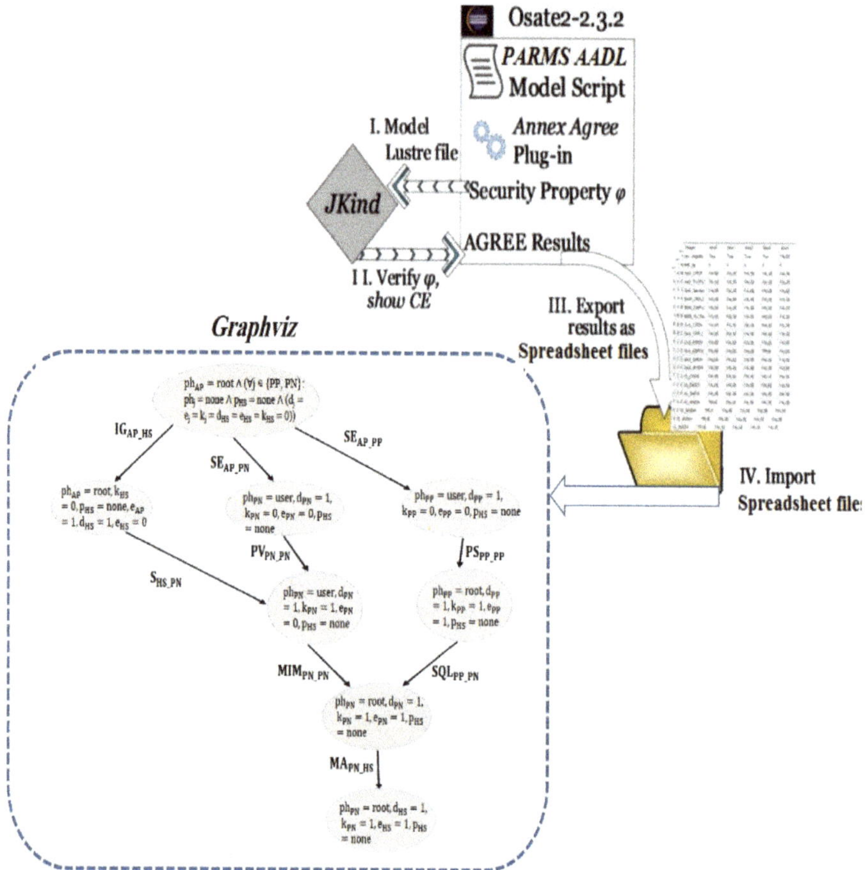

Figure 2. Cyber-attack scenarios implementation workflow. AADL: architecture analysis and design language, CE: counter example.

security property	TRUE	TRUE	TRUE	FALSE
test_e_hs.val	0	0	0	1
thr_ig_aphs.val	TRUE	FALSE	FALSE	FALSE
thr_se_appn.val	FALSE	FALSE	FALSE	FALSE
thr_s_hspn.val	FALSE	TRUE	FALSE	FALSE
thr_se_appn.val	FALSE	FALSE	FALSE	FALSE
thr_ps_pp.val	FALSE	FALSE	FALSE	FALSE
thr_mim_pn.val	FALSE	FALSE	TRUE	FALSE
thr_sql_pppn.val	FALSE	FALSE	FALSE	FALSE
thr_ma_pnhs.val	FALSE	FALSE	FALSE	TRUE
thr_pv_pnpn.val	FALSE	FALSE	FALSE	FALSE

Figure 3. CE1 spreadsheet.

In order to visualize the union of generated cyberattack scenarios (attack graph), the Graphviz tool and DOT graph description language are used. Graphviz is a package of open-source tools used to represent structural information as diagrams of abstract graphs and networks. Graphviz takes the descriptions of graphs in a simple text language [20]. The resulting attack graph shown in Figure 4 consists of arrows and nodes. Each arrow illustrates a possible occurrence of an attack instance, while each node represents the system state resulting from executing the attack instance. An attack scenario is a sequence of attack instances represented by any path from the initial node to the final node in the attack graph. The shown attack graph has three attack scenarios that terminate in a reachable state where the settings of HS can be altered by the attacker. Hence, the attacker can gain a root privilege on the pacemaker, which may threaten the patient's life.

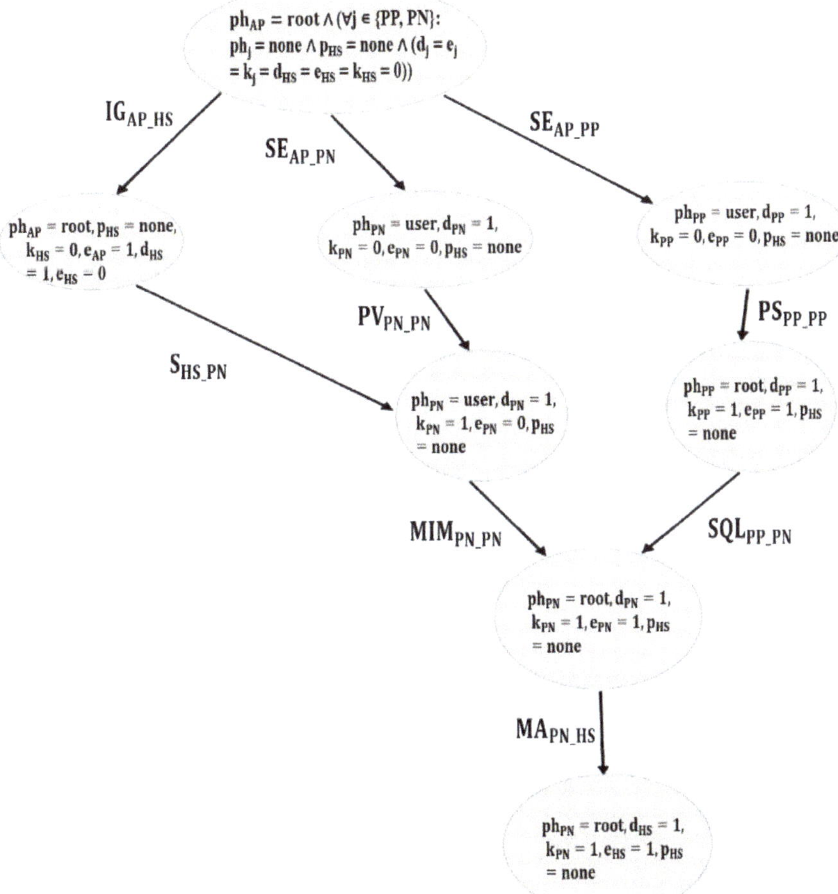

Figure 4. *PARMS* generated attack graph. MA: Malware Injection, SE: Social Engineering, IG: Intelligent Gathering, MiM: Man-in-the-Middle, SQL: SQL Injection, S: Sniffing, PV: Pivoting, PS: Phishing, HS: Home Monitoring Device, PN: Patient Support Network, AP: Access Point, PP: Physician Programmer p_{HS}: Attacker level of privilege on HS, ph_i: Attacker level of privilege on host i, d_i: Data identification of component i; k_i: Confidential data disclosure of component i, e_i: Data alteration of component i.

The generated graph may aid system administrators to decide the placement of appropriate detection and prevention measures. For instance, experimental results showed that an MA attack can never be correctly conducted against the HS without running MIM or SQL attacks first against the PN. Thus, by way of preventing MIM and SQL, the system administrators can also eliminate the MA attack which would immensely enhance the system security.

In addition to that, the MA attack against the HS required exploiting the COTS vulnerability in its operating system or the firmware update vulnerability. Therefore, securing the HS operating system and deploying an intrusion detection system (IDS) between the HS and the PN may prevent the attacker from executing the remaining attacks.

The feasibility of protecting implantable medical devices (IMDs) is explored by [42] without adjusting them by carrying out security strategies completely on an external device called a shield. The shield is placed between the IMD and possible correspondents, e.g., worn on the body close to the implanted device. The shield performs as a gateway that conveys messages between the IMD and accredited endpoints. Such an approach improves the security of IMDs for patients who already have them and enables medical staff to access a protected IMD by discarding the external device or turning it off.

4. Conclusions

In this work, attack graph generation for PARMS is presented using a JKind model checker and DOT language within Graphviz. The idea for modeling is the application of an architectural descriptive language to capture the security-related details of PARMS that an attacker may exploit to impose life-threatening risks to patients. The main goal of this research is to increase the awareness about the security of IoT medical devices. This is done by identifying some of the cyberattacks and estimating their impacts against PARMS. Cyberattacks and vulnerabilities need to be taken into consideration when designing medical IoT devices. Even though some healthcare companies aim to consider within their product development life-cycle safety and security concerns, yet more testing and verification methods are required to produce a systematic method to test the detection or mitigations determined or provided during the safety and security analyses phase against some sophisticated hacking tools. It is important that attacks and defenses be carefully and independently investigated in order to accurately assess risk of the attack and effectiveness of the defense. Determining appropriate detection and mitigation techniques are future directions to pursue.

Author Contributions: Conceptualization, M.I.; methodology, M.I.; software, A.A., and A.M.; validation, M.I., A.A.; formal analysis, M.I.; investigation, M.I., A.A., and A.M; resources, M.I.; data curation, M.I., A.A.; writing—original draft preparation, M.I., A.A., and A.M.; Writing—review and editing, M.I.; visualization, M.I., and A.A.; supervision, M.I.; project administration, M.I.; funding acquisition, M.I. All authors have read and agreed to the published version of the manuscript.

Funding: This research was funded by the Deanship of Graduation Studies and Scientific Research at the German Jordanian University for the seed fund SATS 02/2018.

Acknowledgments: The authors would like to acknowledge Abdulrahman Mhawesh for his valuable discussion on PARMS.

Conflicts of Interest: The authors declare no conflict of interest.

References

1. Abouzakhar, N.S.; Jones, A.; Angelopoulou, O. Internet of Things Security: A Review of Risks and Threats to Healthcare Sector. In Proceedings of the 2017 IEEE International Conference on Internet of Things (iThings) and IEEE Green Computing and Communications (GreenCom) and IEEE Cyber, Physical and Social Computing (CPSCom) and IEEE Smart Data (SmartData), Exeter, UK, 21–23 June 2017; pp. 373–378.
2. Xu, J. Systematic Vulnerability Evaluation of Interoperable Medical Device System using Attack Trees. Master's Theses, Worcester Polytechnic Institute, Worcester, MA, USA, December 2015.

3. Islam, S.M.R.; Kwak, D.; Kabir, H.; Hossain, M.; Kwak, K.S. The Internet of Things for Health Care: A Comprehensive Survey. *IEEE Access* **2015**, *3*, 678–708. [CrossRef]
4. Taylor, C.; Venkatasubramanian, K.; Shue, C.A. Understanding the security of interoperable medical devices using attack graphs. In Proceedings of the 3rd International Conference on Mobile and Ubiquitous Multimedia (MUM '04), College Park, MD, USA, 27–29 October 2014; pp. 31–40.
5. Virat, M.S.; Bindu, S.; Aishwarya, B.; Dhanush, B.; Kounte, M.R.; M, B.S.; N, D.B. Security and Privacy Challenges in Internet of Things. In Proceedings of the 2018 2nd International Conference on Trends in Electronics and Informatics (ICOEI), Tirunelveli, India, 11–12 May 2018; pp. 454–460.
6. Yaqoob, T.; Abbas, H.; Atiquzzaman, M. Security Vulnerabilities, Attacks, Countermeasures, and Regulations of Networked Medical Devices—A Review. *IEEE Commun. Surv. Tutor.* **2019**, in press. [CrossRef]
7. Peck, M. Medical Devices Are Vulnerable to Hacks, But Risk Is Low Overall. Available online: https://spectrum.ieee.org/biomedical/devices/medical-devices-are-vulnerable-to-hacks-but-risk-is-low-overall (accessed on 26 May 2019).
8. Brandom, R. UK Hospitals Hit with Massive Ransomware Attack. Available online: https://www.theverge.com/2017/5/12/15630354/nhs-hospitals-ransomware-hack-wannacry-bitcoin (accessed on 26 May 2019).
9. Center of Internet Security. DDoS Attacks: In the Healthcare Sector. Available online: https://www.cisecurity.org/blog/ddos-attacks-in-the-healthcare-sector/ (accessed on 26 May 2019).
10. Abelson, R.; Goldstein, M. Millions of Anthem Customers Targeted in Cyberattack. Available online: https://www.nytimes.com/2015/02/05/business/hackers-breached-data-of-millions-insurer-says.html (accessed on 26 May 2019).
11. Evenstad, L. NHS Data Breach Caused Details of 150,000 Patients to be Shared. Available online: https://www.computerweekly.com/news/252444145/NHS-data-breach-caused-details-of-150000-patients-to-be-shared (accessed on 26 May 2019).
12. Vincent, J. 1.5 Million Affected by Hack Targeting Singapore's Health Data. Available online: https://www.theverge.com/2018/7/20/17594578/singapore-health-data-hack-sing-health-prime-minister-lee-targeted (accessed on 26 May 2019).
13. Xiao, Y.; Jia, Y.; Cheng, X.; Yu, J.; Liang, Z.; Tian, Z. I Can See Your Brain: Investigating Home-Use Electroencephalography System Security. *IEEE Internet Things J.* **2019**, *6*, 6681–6691. [CrossRef]
14. Luckett, P.; McDonald, J.; Glisson, W. Attack-Graph Threat Modeling Assessment of Ambulatory Medical Devices. In Proceedings of the 50th Hawaii International Conference on System Sciences (2017), Hilton Waikoloa Village, HI, USA, 4–7 January 2017; Volume 4, pp. 3648–3657.
15. Agmon, N.; Shabtai, A.; Puzis, R. Deployment optimization of IoT devices through attack graph analysis. In Proceedings of the 12th Conference on Security and Privacy in Wireless and Mobile Networks (WiSec '19), Miami, FL, USA, 15–17 May 2019; pp. 192–202.
16. Grispos, G.; Glisson, W.B.; Cooper, P. A Bleeding Digital Heart: Identifying Residual Data Generation from Smartphone Applications Interacting with Medical Devices. In Proceedings of the 52nd Hawaii International Conference on System Sciences, Maui, HI, USA, 8–11 January 2019.
17. Ellouze, N.; Rekhis, S.; Boudriga, N.; Allouche, M.; Elouze, N. Cardiac Implantable Medical Devices forensics: Postmortem analysis of lethal attacks scenarios. *Digit. Investig.* **2017**, *21*, 11–30. [CrossRef]
18. Carnegie-Mellon-University. Open Source Aadl Tool Environment for the SAE Architecture. 2018. Available online: http://osate.github.io/index.html (accessed on 15 May 2018).
19. Sheeran, M.; Singh, S.; Stålmarck, G. Checking Safety Properties Using Induction and a SAT-Solver. In *International Conference on Formal Methods in Computer-Aided Design*; Springer: Berlin/Heidelberg, Germany, 2000; Volume 1954, pp. 127–144.
20. Graphviz—Graph Visualization Software. Available online: https://www.graphviz.org/download/ (accessed on 26 May 2019).
21. Sango, M.; Godot, J.; Gonzalez, A.; Nolasco, R.R. Model-Based System, Safety and Security Co-Engineering Method and Toolchain for Medical Devices Design. In Proceedings of the 2019 Design of Medical Devices Conference, Saint Paul, MN, USA, 15–18 April 2019.
22. Tian, Z.; Sun, Y.; Su, S.; Li, M.; Du, X.; Guizani, M. Automated Attack and Defense Framework for 5G Security on Physical and Logical Layers. *arXiv* **2019**, arXiv:1902.04009.

23. Xu, J.; Venkatasubramanian, K.K.; Sfyrla, V. A methodology for systematic attack trees generation for interoperable medical devices. In Proceedings of the 2016 Annual IEEE Systems Conference (SysCon), Orlando, FL, USA, 18–21 April 2016; pp. 1–7.
24. Almohri, H.; Cheng, L.; Yao, D.; Alemzadeh, H. On Threat Modeling and Mitigation of Medical Cyber-Physical Systems. In Proceedings of the 2017 IEEE/ACM International Conference on Connected Health: Applications, Systems and Engineering Technologies (CHASE), Philadelphia, PA, USA, 17–19 July 2017; pp. 114–119.
25. Tu, M.; Spoa-Harty, K.; Xiao, L. Data Loss Prevention Management and Control: Inside Activity Incident Monitoring, Identification, and Tracking in Healthcare Enterprise Environments. *J. Digit. Forensics Secur. Law* **2015**, *10*, 27–44. [CrossRef]
26. Strielkina, A.; Kharchenko, V.; Uzun, D. Availability models for healthcare IoT systems: Classification and research considering attacks on vulnerabilities. In Proceedings of the 2018 IEEE 9th International Conference on Dependable Systems, Services and Technologies (DESSERT), Kyiv, Ukraine, 24–27 May 2018; pp. 58–62.
27. Ghose, N.; Lazos, L.; Rozenblit, J.; Breiger, R. Multimodal graph analysis of cyber attacks. In Proceedings of the IEEE Spring Simulation Conference (SpringSim), Tucson, AZ, USA, 29 April–2 May 2019; pp. 1–12.
28. Ge, M.; Kim, D.S. A Framework for Modeling and Assessing Security of the Internet of Things. In Proceedings of the 2015 IEEE 21st International Conference on Parallel and Distributed Systems (ICPADS), Melbourne, Australia, 14–17 December 2015; pp. 776–781.
29. Kammüller, F.; Tryfonas, T. Formal Modeling and Analysis with Humans in Infrastructures for IoT Health Care Systems. In Proceedings of the Formal Aspects of Component Software, Braga, Portugal, 10–13 October 2017; Volume 10292, pp. 339–352.
30. Doumbouya, M.B.; Kamsu-Foguem, B.; Kenfack, H.; Foguem, C. Combining conceptual graphs and argumentation for aiding in the teleexpertise. *Comput. Boil. Med.* **2015**, *63*, 157–168. [CrossRef] [PubMed]
31. Siddiqi, M.A.; Seepers, R.M.; Hamad, M.; Prevelakis, V.; Strydis, C. Attack-tree-based Threat Modeling of Medical Implants. In Proceedings of the 7th International Workshop on Security Proofs for Embedded Systems, Amsterdam, The Netherlands, 13 September 2018; pp. 32–49.
32. Rios, B.; Butts, J. Security Evaluation of the Implantable Cardiac Device Ecosystem Architecture and Implementation Interdependencies. 2017. Available online: https://a51.nl/whitescope-security-evaluation-implantable-cardiac-device-ecosystem-architecture-and-implementation (accessed on 17 May 2017).
33. Sanders, R.S. The Pulse Generator. In *Cardiac Pacing for the Clinician*; Kusumoto, F.M., Goldschlager, N.F., Eds.; Springer: Boston, MA, USA, 2008; pp. 47–71.
34. Chede, S.; Kulat, K. Design Overview of Processor Based Implantable Pacemaker. *J. Comput.* **2008**, *3*, 49–57. [CrossRef]
35. Ibrahimi, S. *A Secure Communication Model for the Pacemaker a Balance between Security Mechanisms and Emergency Access*; Technische Universiteit Eindhoven: Eindhoven, The Netherland, 2014.
36. Lakshmanadoss, U.; Shah, A.; Daubert, J.P. Telemonitoring of the pacemakers. In *Modern Pacemakers-Present and Future*; IntechOpen: London, UK, 2011; pp. 129–146.
37. Meixell, B.; Forner, E. Out of control: Demonstrating SCADA exploitation. In Proceedings of the Black Hat, Las Vegas, NV, USA, 27 July–1 August 2013.
38. Nagunwa, T. Behind Identity Theft and Fraud in Cyberspace: The Current Landscape of Phishing Vectors. *Int. J. Cyber-Security Digit. Forensics* **2014**, *3*, 72–83. [CrossRef]
39. Patel, N. SQL Injection Attacks: Techniques and Protection Mechanisms. *IJCSE* **2011**, *3*, 199–203.
40. Mebsout, A.; Tinelli, C. Proof certificates for SMT-based model checkers for infinite-state systems. In Proceedings of the Formal Methods in Computer-Aided Design (FMCAD), Mountain View, CA, USA, 3–6 October 2016; pp. 117–124.
41. Halbwachs, N.; Caspi, P.; Raymond, P.; Pilaud, D. The synchronous data flow programming language LUSTRE. *Proc. IEEE* **1991**, *79*, 1305–1320. [CrossRef]
42. Gollakota, S.; Hassanieh, H.; Ransford, B.; Katabi, D.; Fu, K. They can hear your heartbeats: Non-invasive security for implantable medical devices. In Proceedings of the ACM SIGCOMM Conference, Toronto, ON, Canada, 15–19 August 2011; pp. 2–13.

MDPI
St. Alban-Anlage 66
4052 Basel
Switzerland
Tel. +41 61 683 77 34
Fax +41 61 302 89 18
www.mdpi.com

Biosensors Editorial Office
E-mail: biosensors@mdpi.com
www.mdpi.com/journal/biosensors

www.ingramcontent.com/pod-product-compliance
Lightning Source LLC
LaVergne TN
LVHW071001170726
843515LV00006B/1051

* 9 7 8 3 0 3 6 5 7 9 1 2 2 *